AF368913

Impacts of Landscape Change on Water Resources

Impacts of Landscape Change on Water Resources

Editor

Manoj K. Jha

MDPI • Basel • Beijing • Wuhan • Barcelona • Belgrade • Manchester • Tokyo • Cluj • Tianjin

Editor
Manoj K. Jha
North Carolina A&T State
University
USA

Editorial Office
MDPI
St. Alban-Anlage 66
4052 Basel, Switzerland

This is a reprint of articles from the Special Issue published online in the open access journal *Water* (ISSN 2073-4441) (available at: https://www.mdpi.com/journal/water/special_issues/water_landscape_change_impacts_on_water_resources).

For citation purposes, cite each article independently as indicated on the article page online and as indicated below:

LastName, A.A.; LastName, B.B.; LastName, C.C. Article Title. *Journal Name* **Year**, *Article Number*, Page Range.

ISBN 978-3-03943-426-8 (Hbk)
ISBN 978-3-03943-427-5 (PDF)

Contents

About the Editor

Manoj K. Jha is an associate professor in the Department of Civil, Architectural and Environmental Engineering at North Carolina A&T State University. Dr. Jha has about 20 years of experience working in multi-scale modeling of hydrology and water resources/quality, evaluating water management strategies and impact assessment studies due to urban and agriculture best-management practices, land-use change and climate variability. He has made significant contributions in several areas of water resource research nationally and internationally. Specifically, he has developed and applied various fields to watershed scale models to extend the boundaries of our knowledge and understand the impacts of land use and climate change on hydrology, water availability, and water quality. The methods and tools developed by Dr. Jha have made an impact on how we should manage our watersheds to help solve the current water quality problems. Dr. Jha has made significant scientific contributions, as shown by the number of publications in reputable journals and the numerous appearances at professional meetings and the very high and respectable citation index. He has served as a PI/Co-PI on several research grants and contracts worth millions of dollars in funding. His high level of productivity in scientific contributions is reflected by his many awards and honors, including the recent 2019 Senior Researcher of the Year Award in 2019.

Preface to "Impacts of Landscape Change on Water Resources"

In order to manage our valuable water resources, it is imperative to understand the degree of vulnerabilities and resiliency towards changes in the landscape. Continuous changes in land use and land cover can have many drivers, including population growth, urbanization, demand for food, evolution of socio-economic structure, policy regulations, and climate variability. Potential impacts due to these changes could range from changes in water availability (due to changes in losses of water to evapotranspiration and recharge) to degradation in water quality (increased erosion, salinity, chemical loadings, and pathogens). Fields studies are conducted to understand this complexity at local scales, while analyses at regional or watershed scales adopt modeling and simulation strategies. A range of tools, including hydrological, biophysical and ecosystem models, are used (stand-alone or in combination) to investigate important questions regarding impacts in order to inform the decision-making process. These decision analysis tools identify landscape-change impacts, risks, and uncertainties to provide guidance in making key management decisions. In this Special Issue, we include research and discussion topics from field investigations, as well as analytical and modeling studies to better understand the connection between landscape change and water resources at various scales.

Manoj K. Jha
Editor

Editorial

Impacts of Landscape Changes on Water Resources

Manoj K. Jha

Civil, Architectural, and Environmental Engineering, North Carolina A&T State University, Greensboro, NC 27411, USA; mkjha@ncat.edu

Received: 30 July 2020; Accepted: 7 August 2020; Published: 10 August 2020

Abstract: Changes in land use and land cover can have many drivers, including population growth, urbanization, agriculture, demand for food, evolution of socio-economic structure, policy regulations, and climate variability. The impacts of these changes on water resources range from changes in water availability (due to changes in losses of water to evapotranspiration and recharge) to degradation of water quality (increased erosion, salinity, chemical loadings, and pathogens). The impacts are manifested through complex hydro-bio-geo-climate characteristics, which underscore the need for integrated scientific approaches to understand the impacts of landscape change on water resources. Several techniques, such as field studies, long-term monitoring, remote sensing technologies, and advanced modeling studies have been contributing to better understanding the modes and mechanisms by which landscape changes impact water resources. Such research studies can help unlock the complex interconnected influences of landscape on water resources for quantity and quality at multiple spatial and temporal scales. In this Special Issue, we published a set of eight peer-reviewed articles elaborating on some of the specific topics of landscape changes and associated impacts on water resources.

Keywords: landscape change; water resources analysis; water modeling; impact assessment

1. Introduction

Landscape change and its impact on water resources is a vast topic that encompasses fundamental and applied research in multiple dimensions including water resources science and engineering, agriculture, geology, geography, economics, and social sciences. Landscape change can have many manifestations, such as changes due to urbanization, industrialization, commercialization of marginal lands, agriculture, farmers' decisions on the use of croplands, increased use of land through deforestation and drainage of wetlands, government policy decisions for environmental regulations, natural disasters such as floods and droughts, changing climatic and environmental conditions, and others. Subsequently, the impacts of these changes in one form or another on water resources are realized at spatial scales (local impacts contributing to regional scales), temporal scales (short-term vs. long-term changes), changes in water footprints through changes in hydrological processes, and changes in water/environmental quality (sediments, nutrients, and pathogens). For example, changing land use through deforestation and/or drained wetland will influence changes in infiltration and runoff characteristics, thereby affecting evapotranspiration, groundwater recharge, and sediment and water yield [1–4].

It is well-known that the changes in land use have large impacts on water resources; however, quantifying these impacts remains among the more challenging problems in managing water resources [5]. One of the major challenges is the complex interconnection of water within the hydro-bio-geo-climate characteristics [6]. As land use changes, it will alter the water balance through changes in groundwater recharge, runoff, and evapotranspiration. Water movement will be affected due to changes in soil physical properties such as moisture content and soil temperature. Variety of land use types will have associated changes in land characteristics affecting the movement of water in variety of ways. Added to this challenge is the response time of the impacts. For example, groundwater systems response to changes in land use may vary widely from days to decades. Similarly, large scale changes in land use will impact evapotranspiration to a large extent that will be propagated through hydrologic systems and may potentially modify regional weather patterns and climate variability in an unknown future time.

Technological advancements over the years and continuous research efforts have pushed the boundary of science to better understand and quantitatively assess the impacts of landscape change on water resources. While field studies have been proven useful to understand the complexity of impacts at local scales [7], analyses at regional or watershed scales adopt modeling and simulation strategies [8,9]. A range of tools, including hydrological, biophysical, ecosystem models have been developed and used (stand alone or in combination) for investigation and inform the decision-making process. These decision analysis tools identify landscape-change impacts, risks, and uncertainties to provide guidance to make key management decisions [10,11].

This Special Issue presents studies [12–19] that describe the application of a variety of observational and modeling tools and techniques to evaluate the impacts of landscape changes on water resources in watersheds. Landscape changes included change due to agricultural best management practices (BMPs), low impact development (LID) in urban settings, conservation agriculture practices, conversion of erosion hot spot cropland into forest, and others. The impact on water resources included the changes in streamflow, stream and soil temperature, evapotranspiration, sediment, and others. The application of methods included study watersheds in U.S., Hungary, China, South Korea, and Ethiopia.

2. Summary of Papers in the Special Issue

Table 1 shows the comparative analysis of all articles of this Special Issue in terms of the type of the land use change analysis, evaluated impact assessment parameters, and the methods used. Brief summaries of each of the eight published papers are also presented.

Table 1. Comparative analysis of research presented in Special Issue papers.

Landscape Change Analysis	Impact Assessment Parameter(s)	Approach Used	Study Area	Paper
Impact of LID at watershed scale	Surface runoff, subsurface runoff, peak flow, evapotranspiration	Application of an ecohydrological model, Visualizing Ecosystems for Land Management Assessments (VELMA)	East Fork Little Miami River Watershed, OH, USA	[12]
Impact of land use change on stream temperature	Streamflow, stream temperature	Develop a mechanistic stream temperature model in Soil and Water Assessment Tool (SWAT) model	Marys River Watershed, OR, USA	[16]
Impact of open and forested landscapes on soil temperature	Soil temperature	Extending soil temperature model in VELMA	Crest-to-Coast Environmental Monitoring Transect (O'CCMoN) sites, OR, USA	[17]
Impact of landscape on river meandering system	Channel sinuosity vs. forest density and ecological value	GIS analysis of meandering bends between 1952 and 2017 using aerial imagery and UAV (unmanned aerial vehicle)-surveys	Sajó River, Hungary	[18]

Table 1. *Cont.*

Landscape Change Analysis	Impact Assessment Parameter(s)	Approach Used	Study Area	Paper
Impact of optimal placement of BMPs for environmental effectiveness	Streamflow, sediment, BMP cost	Watershed modeling framework using SWAT and optimization algorithm NSGA-II.	Youwuzhen Watershed, China	[13]
Impact of converting erosion hot spot into forested area	Sediment yield	Morgan–Morgan–Finney (DMMF) model	Haean Catchment, South Korea	[14]
Impact of conservation agriculture on a regional scale	Crop yield, water	Agricultural Policy Environmental eXtender (APEX) and GIS-based multi-criteria evaluation (MCE) technique	Ethiopia	[15]
Impact of landscape on vulnerability of flood in an urban watershed	Streamflow, flood extent, inundation	Hydrodynamic model HEC-RAS (Hydrologic Engineering Center—River Analysis System) and hydrologic model SWAT	Blue River, MO, USA	[19]

2.1. Cumulative Effects of Low Impact Development on Watershed Hydrology in a Mixed Land-Cover System, by Hoghooghi et al., 2018

LID practices are designed to reduce the impact of land use change on hydrology. This study used a spatially explicit ecohydrological model VELMA, to assess the impact of LID techniques at the watershed scale. The authors calibrated and validated the model for streamflow. Hydrological effects of three common LID practices (rain gardens, permeable pavement, and riparian buffers) were tested on a 0.94 km^2 mixed land cover semi-urban watershed (27 percent impervious) in Ohio, USA. LID practices were shows to perform as expected for effectively reducing the peak flow and increasing infiltration but with limited efficiency in semi-urban watershed.

2.2. Modeling Landscape Change Effects on Stream Temperature Using the Soil and Water Assessment Tool, by Mustafa et al., 2018

Stream temperature is an important factor in regulating fish behavior and habitat. This study investigates the impact of landscape change on stream temperature. The authors developed a mechanistic stream temperature module within the watershed modeling environment of the SWAT model and applied it to the 782 km^2 watershed in Oregon, USA. The model was calibrated for flow and stream temperature before examining the changes in stream temperature due to change in land use. The model was able to capture the increased stream temperatures in agricultural sub-basins compared with forested sub-basins.

2.3. Improved Soil Temperature Modeling Using Spatially Explicit Solar Energy Drivers, by Halama et al., 2018

Soil temperature affects ecosystem properties including increasing the water temperature. This study demonstrated that local solar energy information improved soil temperature modeling estimates simulated by a soil temperature subroutine within a larger ecohydrological watershed model VELMA. Authors calibrated the model using the data available from Oregon's Crest-to-Coast Environmental Monitoring Transect (O'CCMoN) sites. Results demonstrated the benefit of including spatially explicit representations of solar energy within watershed-scale models that simulate soil temperature.

2.4. Issues of Meander Development: Land Degradation or Ecological Value? The Example of the Sajó River, Hungary, by Bertalan et al., 2018

River channels and their surrounding floodplains enhance landscape evolution and the diversification of environments. This study investigates the geomorphological development and effects of bank erosion along meandering Sajó River in Hungary. Authors performed GIS analysis of three consecutive meandering bends over 10 periods between 1952 and 2017 based on archive aerial imagery and UAV-surveys. Analyses revealed that the meandering (channel sinuosity) was directly proportional to forest density (dominant, compact, and connected) which provided high ecological value.

2.5. Effects of Different Spatial Configuration Units for the Spatial Optimization of Watershed Best Management Practice Scenarios, by Zhu et al., 2019

Variation in spatial configurations of BMPs at the watershed scale may have significantly different environmental effectiveness. This study investigated and compared the effects of four main types of spatial configuration units for BMP scenarios optimization. Optimization was conducted based on a fully distributed watershed modeling framework, the Spatially Explicit Integrated Modeling System (SEIMS) using SWAT, and an intelligent optimization algorithm Non-dominated Sorting Genetic Algorithm II (NSGA-II). Results showed that the different BMP configuration yielded significant differences in near-optimal Pareto solutions, optimizing efficiency, and spatial distribution of BMP scenarios. BMP configuration units that support the adoption of expert knowledge on the spatial relationships between BMPs and spatial locations (e.g., hydrologically connected fields, slope position units) are considered to be the most valuable spatial configuration units for watershed BMP scenarios optimization and integrated watershed management.

2.6. Evaluating the Effectiveness of Spatially Reconfiguring Erosion Hot Spots to Reduce Stream Sediment Load in an Upland Agricultural Catchment of South Korea, by Choi et al., 2019

Soil erosion has a negative impact on the environment and socioeconomic factors by degrading the quality of both nutrient-rich surface soil and water. This modeling study demonstrated the effectiveness of converting soil erosion hot spots within the watershed into forest for reducing the sediment yield significantly.

2.7. Scaling-Up Conservation Agriculture Production System with Drip Irrigation by Integrating MCE Technique and the APEX Model, by Assefa et al., 2019

Conservation agriculture, which promotes no-till, mulching, and diverse cropping, provides higher water use efficiency in addition to improving soil fertility and crop yield. This study demonstrated the scaling-up impacts of conservation agriculture on a regional scale. The calibrated biophysical model APEX in combination with GIS-based multi-criteria evaluation (MCE) technique was used to extend the modeling analysis to the national scale in Ethiopia. Results indicated that the conservation agriculture with drip irrigation technology could improve groundwater potential for irrigation up to five folds and intensify crop productivity by up to three to four folds across the nation.

2.8. Flooding Urban Landscapes: Analysis Using Combined Hydrodynamic and Hydrologic Modeling Approaches, by Jha and Afreen, 2020

Urban landscape dictates the extent of inundation during a flood event, affecting vulnerable infrastructures. This study presents a systematic approach of combining hydrodynamic model HEC-RAS with hydrologic model SWAT in delineating flood inundation zones, and subsequently assessing the vulnerability of critical infrastructures in the Blue River Watershed in Kansas City, Missouri. Results demonstrate the usefulness of such combined modeling systems to predict the extent of flood inundation and thus support analyses of management strategies to deal with the risks associated with critical infrastructures in an urban setting.

3. Conclusions

Landscape changes have direct linkages with changes in hydrology in terms of water balance components. As land use characteristics change, it will alter the hydrology at the local scale, leading to the impacts on water availability and associated water quality to regional scales and at various temporal scales. The papers in this Special Issue describe the applications of a variety of observational and modeling tools and techniques to evaluate the impacts of landscape changes on water resources. The studies can be categorized into four subject areas: (1) impact assessment due to implementation of management practices [12–15], (2) impact of landscape on stream and soil temperature [16,17], (3) landscape and river meandering [18], and (4) landscape for flood inundation [19].

Funding: This received no external funding.

Conflicts of Interest: The author declares no conflict of interest.

References

1. Vandas, S.J.; Winter, T.C.; Battaglin, W.A. Water and the Environment. American Geosciences Institute (AGI) Environmental Awareness Series. 2002. ISBN 0-922152-63-2. Available online: http://www.agiweb.org/environment/publications/water.pdf (accessed on 25 July 2020).
2. Kling, C.L.; Panagopoulos, Y.; Rabotyagov, S.S.; Valcu, A.M.; Gassman, P.W.; Campbell, T.; White, M.J.; Arnold, J.; Srinivasan, R.; Jha, M.K.; et al. LUMINATE: Linking agricultural land use, local water quality and Gulf of Mexico hypoxia. *Eur. Rev. Agric. Econ.* **2014**, *41*, 431–459. [CrossRef]
3. Jha, M.K.; Schilling, K.E.; Gassman, P.W.; Wolter, C.F. Targeting land-use change for nitrate-nitrogen load reductions in an agricultural watershed. *J. Soil Water Conserv.* **2010**, *65*, 342–352. [CrossRef]
4. Rabotyagov, S.; Jha, M.K.; Campbell, T. Impact of crop rotations on optimal selection of conservation practices for water quality protection. *J. Soil Water Conserv.* **2010**, *65*, 369–380. [CrossRef]
5. Stonestrom, D.A.; Scanlon, B.R.; Zhang, L. Introduction to special section on impacts of land use change on water resources. *Water Resour. Res.* **2009**, *45*, W00A00. [CrossRef]
6. Panagopoulos, Y.; Gassman, P.W.; Jha, M.K.; Kling, C.L.; Campbell, T.; Srinivasan, R.; White, M.; Arnold, J.G. A refined regional modeling approach for the Corn Belt—Experiences and recommendations for large-scale integrated modeling. *J. Hydrol.* **2015**, *524*, 348–366. [CrossRef]
7. Assefa, T.T.; Jha, M.K.; Reyes, M.R.; Tilahun, S.A.; Worqlul, A. Experimental evaluation of conservation agriculture with drip irrigation for water productivity in sub-Saharan Africa. *Water* **2019**, *11*, 530. [CrossRef]
8. Assefa, T.T.; Jha, M.K.; Reyes, M.R.; Worqlul, A. Modeling the impacts of conservation agriculture with drip irrigation system on hydrology and water management in sub-Sahara Africa. *Sustainability* **2018**, *10*, 4763. [CrossRef]
9. Jha, M.K. Evaluating Hydrologic Response of an Agricultural Watershed for Watershed Analysis. *Water* **2011**, *3*, 604–617. [CrossRef]
10. Schilling, K.E.; Gassman, P.W.; Kling, C.L.; Campbell, T.; Jha, M.K.; Wolter, C.F.; Arnold, J.G. The potential for agricultural land use change to reduce flood risk in a large watershed. *Hydrol. Process.* **2013**, *28*, 3314–3325. [CrossRef]
11. Jha, M.K.; Gassman, P.W.; Arnold, J.G. Water quality modeling for the Raccoon River Watershed Using SWAT. *Trans. ASABE* **2007**, *50*, 479–493. [CrossRef]
12. Hoghooghi, N.; Golden, H.; Bledsoe, B.; Barnhart, B.; Brookes, A.; Djang, K.; Halama, J.; McKane, R.; Nietch, C.; Pettus, P. Cumulative effects of low impact development on watershed hydrology in a mixed land-cover system. *Water* **2018**, *10*, 991. [CrossRef] [PubMed]
13. Zhu, L.; Qin, C.; Zhu, A.; Liu, J.; Wu, H. Effects of different spatial configuration units for the spatial optimization of watershed best management practice scenarios. *Water* **2019**, *11*, 262. [CrossRef]
14. Choi, K.; Maharjan, G.; Reineking, B. Evaluating the effectiveness of spatially reconfiguring erosion hot spots to reduce stream sediment load in an upland agricultural catchment of South Korea. *Water* **2019**, *11*, 957. [CrossRef]
15. Assefa, T.; Jha, M.; Worqlul, A.; Reyes, M.; Tilahun, S. Scaling-up conservation agriculture production aystem with drip irrigation by integrating MCE technique and the APEX model. *Water* **2019**, *11*, 2007. [CrossRef]

16. Mustafa, M.; Barnhart, B.; Babbar-Sebens, M.; Ficklin, D. Modeling landscape change effects on stream temperature using the Soil and Water Assessment Tool. *Water* **2018**, *10*, 1143. [CrossRef]

17. Halama, J.; Barnhart, B.; Kennedy, R.; McKane, R.; Graham, J.; Pettus, P.; Brookes, A.; Djang, K.; Waschmann, R. Improved soil temperature modeling using spatially explicit solar energy drivers. *Water* **2018**, *10*, 1398. [CrossRef]

18. Bertalan, L.; Novák, T.; Németh, Z.; Rodrigo-Comino, J.; Kertész, Á.; Szabó, S. Issues of meander development: Land degradation or ecological value? The example of the Sajó River, Hungary. *Water* **2018**, *10*, 1613. [CrossRef]

19. Jha, M.; Afreen, S. Flooding urban landscapes: Analysis using combined hydrodynamic and hydrologic modeling approaches. *Water* **2020**, *12*, 1986. [CrossRef]

Article

Cumulative Effects of Low Impact Development on Watershed Hydrology in a Mixed Land-Cover System

Nahal Hoghooghi [1,2], **Heather E. Golden** [3,*], **Brian P. Bledsoe** [2], **Bradley L. Barnhart** [4], **Allen F. Brookes** [4], **Kevin S. Djang** [5], **Jonathan J. Halama** [4], **Robert B. McKane** [4], **Christopher T. Nietch** [6] **and Paul P. Pettus** [4]

[1] Oak Ridge Institute for Science and Education, c/o US Environmental Protection Agency, Office of Research and Development, National Exposure Research Laboratory, Cincinnati, OH 45268, USA; nahalh@uga.edu

[2] Institute for Resilient Infrastructure Systems, College of Engineering, University of Georgia, Athens, GA 30602, USA; bbledsoe@uga.edu

[3] National Exposure Research Laboratory, Office of Research and Development, US Environmental Protection Agency, Cincinnati, OH 45268, USA

[4] Western Ecology Division, National Health and Environmental Effects Research Laboratory, US Environmental Protection Agency, Corvallis, OR 97330, USA; Barnhart.Brad@epa.gov (B.L.B); Brookes.Allen@epa.gov (A.F.B.); Halama.Jonathan@epa.gov (J.J.H.); Mckane.Bob@epa.gov (R.B.M.); Pettus.Paul@epa.gov (P.P.P.)

[5] Inoventures LLC, Western Ecology Division, National Health and Environmental Effects Research Laboratory, c/o US Environmental Protection Agency, Corvallis, OR 97330, USA; Djang.Kevin@epa.gov

[6] National Risk Management Research Laboratory, US Environmental Protection Agency, Cincinnati, OH 45268, USA; Nietch.Christopher@epa.gov

* Correspondence: Golden.Heather@epa.gov; Tel.: +1-513-569-7773

Received: 20 June 2018; Accepted: 20 July 2018; Published: 27 July 2018

Abstract: Low Impact Development (LID) is an alternative to conventional urban stormwater management practices, which aims at mitigating the impacts of urbanization on water quantity and quality. Plot and local scale studies provide evidence of LID effectiveness; however, little is known about the overall watershed scale influence of LID practices. This is particularly true in watersheds with a land cover that is more diverse than that of urban or suburban classifications alone. We address this watershed-scale gap by assessing the effects of three common LID practices (rain gardens, permeable pavement, and riparian buffers) on the hydrology of a 0.94 km^2 mixed land cover watershed. We used a spatially-explicit ecohydrological model, called Visualizing Ecosystems for Land Management Assessments (VELMA), to compare changes in watershed hydrologic responses before and after the implementation of LID practices. For the LID scenarios, we examined different spatial configurations, using 25%, 50%, 75% and 100% implementation extents, to convert sidewalks into rain gardens, and parking lots and driveways into permeable pavement. We further applied 20 m and 40 m riparian buffers along streams that were adjacent to agricultural land cover. The results showed overall increases in shallow subsurface runoff and infiltration, as well as evapotranspiration, and decreases in peak flows and surface runoff across all types and configurations of LID. Among individual LID practices, rain gardens had the greatest influence on each component of the overall watershed water balance. As anticipated, the combination of LID practices at the highest implementation level resulted in the most substantial changes to the overall watershed hydrology. It is notable that all hydrological changes from the LID implementation, ranging from 0.01 to 0.06 km^2 across the study watershed, were modest, which suggests a potentially limited efficacy of LID practices in mixed land cover watersheds.

Keywords: LID practices; watershed scale; impervious area; peak flow; surface runoff; shallow subsurface runoff and infiltration; evapotranspiration

---

1. Introduction

Urbanization alters natural hydrological systems by altering stream channel networks (e.g., channelization and burial), creating microclimates (e.g., urban heat islands), and generating rapid runoff from precipitation and snowmelt events [1]. These changes have direct impacts on surface and groundwater quantity and quality. Conventional urban stormwater management practices are often developed to control runoff and minimize flooding; however, these systems can be costly and may not directly address issues, such as reductions in infiltration and groundwater storage via impervious surfaces that may lead to urban flooding, erosion, and the degradation of water quality [2].

In recent years, alternative stormwater management practices, such as Low Impact Development (LID), have been adopted (e.g., bioretention cells or rain gardens, permeable pavements, and bioswales) in many urban and suburban areas [3,4]. LID, also called sustainable urban drainage systems (SUDSs), among other globally varying names [5], is an approach that uses soils, vegetation, and landscape design to control nonpoint source runoff and pollutants in urban systems. A goal of LID is to promote watershed resilience through "green" design [6].

There is a growing body of literature focused on evaluating the local (e.g., plot or site) scale effectiveness of LID. Several recent papers have synthesized the key findings of studies assessing the effects of different LID practices, including field experiments and modeling studies, on water quantity (e.g., peak flow and runoff volume) and water quality (e.g., nitrogen, phosphorous, and total suspended solids) at local scales [7–11]. These previous studies provide foundational research for scaling LID approaches to watersheds. However, limited evidence of LID effectiveness at the watershed scale exists [12,13], and research focusing on LID impacts at watershed scales is just beginning to emerge [14]. Therefore, questions remain about how LID practices can individually or cumulatively affect watershed hydrology [14,15].

Experimental studies, designed to investigate the watershed-scale effects of LID, have provided critical insights into how watershed hydrology responds to these approaches. For example, Jarden et al. [16] designed a paired watershed approach to quantify the effect of street-connected bioretention cells, rain gardens, and rain barrels on peak discharge and total storm runoff. The results from the subwatershed with smaller LID lots and underdrain connections showed a substantial reduction in peak discharge (up to 33%) and total storm runoff (up to 40%). Additionally, recent field-based research provides evidence of the cumulative watershed scale effects of LID on hydrologic responses, such as peak flows and pollutant loads [17–20]. Such experimental studies can be resource intensive (e.g., financial, personnel, time) [21]; however, process-based models provide a means to go beyond measured data and explore the projected "what if" LID scenarios using potentially less resources.

Process-based or mechanistic watershed models, which simulate hydrological (and other) processes and outputs for different water balance components (e.g., streamflow, evapotranspiration), are critical tools to understand the influence of LID practices on watershed processes [22]. The Storm Water Management Model (SWMM) [23] has been used to simulate the effects of different LID practice implementations (porous pavement, rain barrels, and rain garden) on runoff and flood risk reductions in an 87.6 km^2 urban watershed and model the cumulative effects of street-side bioretention cells, rain gardens, and rain barrels in a 0.12 km^2 residential watershed [21]. Overall, the results indicate increases in evaporation and infiltration, as well as decreases in surface runoff and discharge, across different return periods. The performance of LID practices (rain gardens, permeable pavements, and rainwater harvesting tanks) has also been evaluated under different urban land use densities using the Soil and Water Assessment Tool (SWAT) [24], demonstrating that the effectiveness of LID practices differs among the urban land use densities [25].

The aforementioned studies and others (e.g., [26–29]) advance current knowledge on the effectiveness of LID practices at watershed scales using various process-based model approaches; however, all of these studies focus on watersheds that are entirely urban or suburban. A clear need exists for an understanding of the extent to which LID approaches are effective in mixed land cover watersheds, i.e., those with urban and suburban land cover *in addition to others* (e.g., forest and agriculture). Furthermore, a spatially explicit approach toward representing LID practices and associated hydrological processes to analyze the effects of varying patterns of mixed land use and land cover under different management practices is critical [30], as most approaches are challenged with representing spatial landscape heterogeneities [31].

In this paper, we assess how LID implementation affects watershed hydrologic responses in a mixed land cover watershed. Specifically, we ask: How does the type and extent of LID practices affect water balance components, including surface runoff, peak flows, evapotranspiration, shallow subsurface flow, and infiltration, in a mixed land cover watershed? We do this by using a spatially-explicit ecohydrological model, called Visualizing Ecosystems for Land Management Assessments (VELMA) [32] for a variety of scenarios associated with LID and the implementation of forested riparian buffers. Our study is one of the first, to our knowledge, to examine LID implementation at the watershed scale using spatially explicit modeling approaches in a system with mixed suburban, agricultural and forest land cover. As a result, we discuss the implications of this study for effective stormwater management in mixed land cover systems and future research directions toward this goal.

2. Materials and Methods

2.1. Study Area Description

The Shayler Crossing (SHC) watershed is a subwatershed of the East Fork Little Miami River Watershed in southwest Ohio, USA and falls within the Till Plains region of the Central Lowland physiographic province. The Till Plains region is a topographically young and extensive flat plain, with many areas remaining undissected by even the smallest stream. The bedrock is buried under a mantle of glacial drift 3–15 m thick [33,34]. The Digital Elevation Model (DEM) has a maximum value of ~269 m (North American_1983 datum) within the watershed boundary (Figure 1). The soils are primarily the Avonburg and Rossmoyne series, with high silty clay loam content and poor to moderate infiltration [35]. Average annual precipitation for the period, 1990 through 2011, was 1097.4 ± 173.5 mm. Average annual air temperature for the same period was 12 °C [36].

We considered SHC a mixed land cover watershed, located on the east side of Cincinnati, Ohio, with a drainage area of 0.92 km^2 (Figure 1). The primary land uses consist of 64.1% urban or developed area (including 37% lawn, 12% building, 6.5% street, 6.4% sidewalk, and 2.1% parking lot and driveway), 23% agriculture, and 13% deciduous forest (Table 1). Total imperviousness covers approximately 27% of the watershed area, the majority of which is directly connected to a storm sewer system without any intermediary controls [30]. The watershed was chosen for this study because it is part of the East Fork Little Miami River Watershed, where a long-term monitoring and focused modeling effort is being conducted by the US Environmental Protection Agency (EPA), Office of Research and Development (ORD), Ohio Environmental Protection Agency (Ohio EPA), and Clermont County (Ohio) Stormwater Division.

2.2. Input Data

We obtained average daily precipitation and temperature data from a weather station, located approximately 13 km from the north boundary of the watershed at 84.2909° W, 39.194° N [37]. Streamflow has been monitored, from 3 April 2006 to the present day, using a stage sensor (600 LS Sonde with temperature, conductivity, and shallow vented level sensors, YSI Inc., Yellow Springs, OH, USA) at the watershed outlet. Water depth was recorded at 10-min intervals and converted to streamflow (m^3 s^{-1}) using a rating curve, developed by US EPA. We obtained a 10 m resolution DEM, Soil Survey Spatial Tabular (SSURGO 2.2) soil data, and National Land Cover Dataset (NLCD) land use data from the Natural Resources Conservation Service (NRCS) Geospatial Data Gateway [38]. We further used an impervious area shape file from Clermont County, Ohio through the Center for Urban Green Infrastructure Engineering, Inc. (Milford, OH, USA).

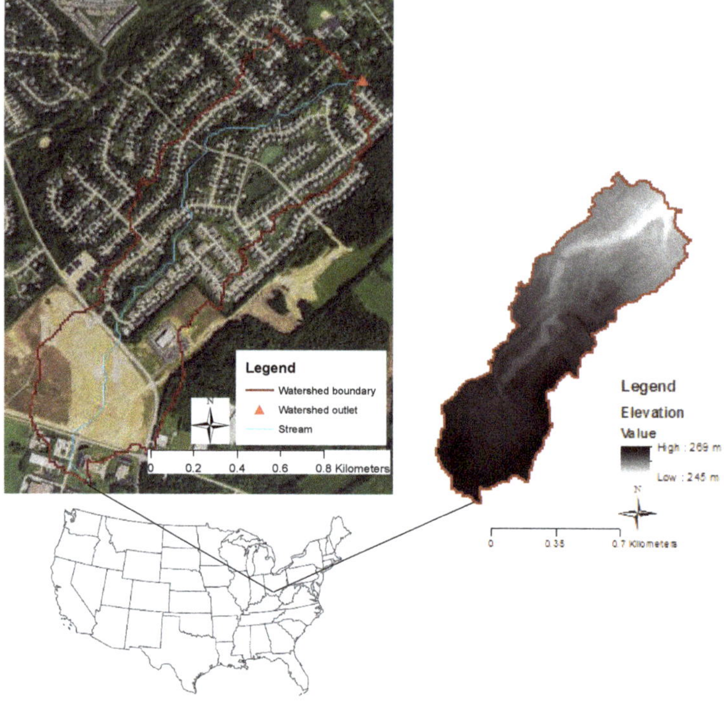

Figure 1. The study area is a 0.92 km^2 subwatershed (Shayler Crossing) of the East Fork Little Miami River watershed, located on the east side of Cincinnati, Ohio in Clermont County, USA. The watershed outlet is identified as a red triangle.

Table 1. Summary of Shayler Crossing watershed land use area and characteristics.

Land Use Type	Area (km^2)	% of Watershed	% of Total Imperviousness
Lawn	0.34	37.0	-
Agriculture (corn)	0.21	23.0	-
Forest	0.12	13.0	-
Building	0.11	12.0	44.5
Street	0.06	6.5	24.0
Sidewalk	0.06	6.4	23.5
Parking Lot and Driveway	0.02	2.1	8.0

2.3. Model Description

To simulate the effect of LID on watershed hydrology, we used the Visualizing Ecosystems for Land Management Assessments (VELMA) model. VELMA is a spatially distributed ecohydrological model that couples watershed hydrology and carbon (C) and nitrogen (N) cycling in plants and soils, and the transport of water, C, and N from the terrestrial landscape to streams [32]. VELMA is not an "urban hydrology" model according to the strict tradition of stormwater management models (e.g., SWMM). Its key strengths are its spatially explicit representation of hydrological and biogeochemical processes and broad applicability to a variety of ecosystems, such as forest, agricultural, and urban, in order to assess the effects of LID in mixed land cover systems. Urban LID practices can be represented in the model using modifications to present watershed permeability, lateral and horizontal hydraulic conductivities, and land cover (see Section 2.5). VELMA's spatially explicit grid-based structure affords the capacity to represent transitions from directly connected to indirectly connected impervious areas by replacing values on a cell by the cell basis for the aforementioned model representations. The model is also capable of scaling hydrologic and biogeochemistry responses across multiple spatial (hillslopes to basins) and temporal (days to centuries) scales [21]. VELMA's visualization and interactivity features are packaged in an open-source, open-platform programming environment (Java/Eclipse) [32].

VELMA's modeling domain is a three-dimensional matrix that includes information regarding surface topography, land use, and four soil layers. VELMA uses a distributed soil column framework to model the lateral and vertical movement of water and nutrients through the four soil layers. A soil water balance is solved for each layer. The soil column model has three coupled submodels: (1) A hydrological model that simulates the vertical and lateral movement of water within the soil and losses of water from soil and vegetation in the atmosphere; (2) a soil temperature model that simulates daily soil layer temperatures based on surface air temperature; and (3) a biogeochemistry model that simulates C and N dynamics.

A simple logistical function, based on the degree of saturation, is applied to capture the breakthrough characteristic of soil water. Potential evapotranspiration (PET) is estimated using the simple temperature-based method of Hamon [39]. Evapotranspiration (ET) increases exponentially as soil water storage increases, and it reaches the PET rate as the soil water storage reaches saturation. The VELMA simulator engine allows for the specification of a spatial data map, with permeability fractions for each grid cell value (here, each 10 m grid cell). The grid's permeability fractions are taken into account when determining how much of a cell's total water inflow (e.g., from rain, snow melt, and lateral surface movement) penetrates into the first layer of the soil column. A permeability of 0 is completely impermeable (no water penetrates from the surface to the first soil layer), and 1 is completely permeable (all water penetrates from the surface to the first soil layer).

The soil column model is placed within a watershed framework to create a spatially distributed model applicable to watersheds (Figure 2, shown here with LID practices). Adjacent soil columns interact through down-gradient water transport. Water entering each pixel (via precipitation or flow from an adjacent pixel) can either first infiltrate into the implemented LID and the top soil layer, and then to the downslope pixel, or continue its downslope movement as the lateral surface flow. Surface and subsurface lateral flow are routed using a multiple flow direction method, as described in Abdelnour et al. [21]. A detailed description of the processes and equations can be found in McKane et al. [32], Abdelnour et al. [21], Abdelnour et al. [40].

2.4. Watershed Model Setup

We used VELMA's pre-processor tool (called Java Processing Digital Elevation Model (JPDEM) [32,41]) to fill sinks, determine flow direction, and compute the flow contribution area of a 10 m DEM [32]. The watershed boundary was delineated, and the watershed outlet was assigned using VELMA's pre-processor [32]. All DEM, soil, and land use maps were clipped so that they have the same number of columns and rows for the American Standard Code for Information Interchange (ASCII grid, Esri, Inc., Esri grid format ArcGIS Desktop 10.0 Help, http://desktop.arcgis.com/en/arcmap/10.3/manage-data/raster-and-images/esri-grid-format.htm) input in VELMA. The soil and the land use maps contained ID numbers for every cell in the simulation area, which corresponded to one or more of VELMA's simulator configurations. We assigned two of VELMA's soil configurations to represent Rossymoyne and Avonburg soil types and seven land use configurations to represent agriculture, forest, lawn, buildings, streets, sidewalks, parking lots and driveways. We merged wet pond pixels with lawn pixels because currently lakes and ponds are not implementable in VELMA.

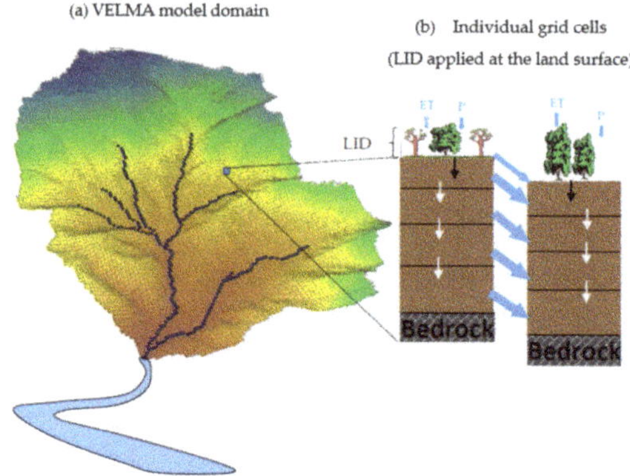

Figure 2. Generalized structure of the VELMA model, and applications for LID. VELMA's domain is a watershed (**a**). Each grid cell within the watershed has four soil layers (**b**), including LID applied to the land surface). Infiltration from LID is transported to the first soil layer (black arrow). Further vertical transport of water (in this paper), carbon, and nitrogen transport can occur between each grid cell's four layers (white arrows), and surplus water (or carbon, nitrogen) is transferred from a grid cell to the adjacent, most down gradient cell(s) in the watershed (blue arrows). P = Precipitation; ET = evapotranspiration. Modified from Abdelnour et al. [21].

2.5. Base Model Parameterization, Calibration and Validation

We performed the base model calibration with daily streamflow at the outlet of the watershed from 1 January 2009 to 31 December 2010 with 2008 as a model warmup period and from 1 January 2011 to 31 December 2011 as a validation period. Our calibration period (2009 and 2010) included normal precipitation years (1040 and 1046 mm), and our verification period (2011) was a wet year (1660 mm). We defined a 'wet' period as greater than one standard deviation from the mean precipitation (>1270.1 mm) and a dry period as less than one standard deviation (<923.9 mm).

Calibration was conducted through both semi-automatic and manual calibrations. We used autocalibration to screen for sensitive parameters and reduce the solution space. Manual calibration was implemented as a second phase to further refine the parameter values. For the initial automatic calibration, we used the MOEA-VELMA calibration tool that links VELMA with the Multiobjective Evolutionary Algorithm (MOEA) [42] framework in Java. The MOEA framework uses evolutionary algorithms to solve multiobjective optimization problems, and the MOEA-VELMA calibration tool leverages this ability to tune model input parameters to minimize the differences between simulated results and observed data. Several parameters were chosen to calibrate the model, including soil layer thickness, saturated hydraulic conductivity, porosity fraction, bulk density, wilting point, field capacity, and PET parameters. The MOEA-VELMA calibration tool then implemented NSGA-II [43], using the MOEA framework, and searched for the optimal set of input parameters to optimize our objective function, that is, Nash Sutcliffe Efficiency (NSE) [44] for the observed and predicted daily streamflow:

$$NSE = 1 - \frac{\sum_{i=1}^{n}|O_i - P_i|^2}{\sum_{i=1}^{n}|O_i - \overline{O}|^2} \tag{1}$$

where O_i is the ith measured variable (e.g., discharge), P_i is the ith predicted variable, $\overline{O}$ is the arithmetic average of the measured variable, and n is the total number of observations. The NSE coefficient ranges between 1 (perfect fit) and negative infinity. An efficiency below zero implies that the mean value of the observed value is a better predictor than the model.

After almost 500 simulations, we narrowed the range of selected sensitive parameters and ran the MOEA-VELMA calibration tool for an additional 500 simulations. Then, we picked the solutions with a higher NSE and used those parameter ranges in the manual calibration.

After the initial semi-automatic calibration, we conducted manual calibration through visual analysis to capture trends in observed streamflow, using NSE in addition to percent bias (*PBIAS*) [45] and root mean squared error (RMSE) [46]. PBIAS measures the average tendency of the predicted data to be larger or smaller than observed values. It is also measures over- and underestimation of bias [44]:

$$PBIAS = \frac{\sum_{i=1}^{n}(O_i - P_i)}{\sum_{i=1}^{n} O_i} \times 100 \tag{2}$$

and RMSE is the square root of the mean square error and varies from zero to large positive values:

$$RMSE = \sqrt{\frac{1}{n} \sum_{i=1}^{n}(O_i - P_i)^2} \tag{3}$$

To ensure the simulations provided reasonable volumetric matches with observed data, we also used a total simulated to total observed annual streamflow ratio (Sim:Obs) for each simulation year. If the Sim:Obs was >1, simulated streamflow from the year exceeded that of the observed streamflow. If it was <1, the opposite was true, and Sim:Obs = 1 suggested a perfect match between the total annual simulated and observed streamflow.

The soil thickness of each layer was parameterized using United States Department of Agriculture (USDA) soil survey data for the study area [35]. We used the MOEA-VELMA calibration tool to calibrate saturated vertical and horizontal hydraulic conductivities for each soil layer. Other soil physical characteristics (porosity, field capacity, wilting point, and bulk density) were obtained based on soil texture class (Table 2) [32]. We obtained the first term of the PET *Hamon* equation (petParam1) for different cover types using the MOEA-VELMA calibration tool, with the second term of *Hamon* equation set to a constant value of 0.622, based on Abdelnour et al. [21]. A *be* parameter is a calibration constant; it is an ET coefficient used in the logistic equation that computes ET from PET. We estimated this parameter value from autocalibration. Air density (*roair*) was constant and set to 1300 g m^{-3} (Table 3). We adjusted all soil physical characteristics and PET parameters to best match the observed streamflow during manual calibration. The parameters and their final model values are shown in Tables 2 and 3. Setting soil parameters to zero produces an error in the VELMA output; therefore, we set soil parameter values for impervious areas to those of the clay soil texture class, using the approach by McKane et al. [32].

Table 2. Soil parameters for the base model (* = calibrated; all other values from McKane et al. [32]). *R* stands for Rossmoyne and *A* stands for Avonburg soil type.

Parameter	Description	Soil Type	Layer	Value	Unit
z	Soil layer thickness	All	1, 2, 3, 4	500, 500, 12,000, 12,000	mm
$K_{s,l}$ *	Saturated lateral hydraulic conductivity	*R* and *A* Impervious area	1, 2, 3, 4	130, 100, 80, 30 50, 30, 15, 10	-
$K_{s,v}$ *	Saturated vertical hydraulic conductivity	*R* and *A* Impervious area	1, 2, 3, 4	14, 14, 10, 10 7, 7, 5, 5	-
n	Porosity fraction	*R* and *A* Impervious area	All	0.501 0.475	-
P_b	Bulk density	*R* and *A* Impervious area	All	1.42 1.21	g cm^{-3}
θ_{wp}	Wilting point	*R* and *A* Impervious area	All	0.133 0.272	-
θ_{fc}	Field capacity	*R* and *A* Impervious area	All	0.33 0.396	-

Table 3. Calibrated Potential Evapotranspiration parameters for the base model.

Parameter	Description	Cover Type	Value	Unit
PetParam1	First term of PET *Hamon* equation	Agriculture (corn) Forest Lawn Impervious area	0.20 0.30 0.15 0.05	-
PetParam2	Second term of PET *Hamon* equation	Agriculture (corn) Forest Lawn Impervious area	0.622	-
TemperaturePetOff	PET is only active when air temperature is greater than this value	Agriculture (corn) Forest Lawn Impervious area	-3	C°
roair	Air density	Agriculture (corn) Forest Lawn Impervious area	1300	$g\,m^{-3}$
be	ET coefficient used in the logistic equation that computes ET from PET	Agriculture (corn) Forest Lawn Impervious area	3.07	-
noTranspirationPetFraction	The fraction of PET available outside of this cover's growing season	Agriculture (corn) Forest Lawn Impervious area	1	-

2.6. Low Impact Development (LID) Configurations, Scenarios, and Model Parameters

To evaluate the effectiveness of LID practices based on the relative daily changes in watershed hydrology compared to the calibrated base model, we simulated three types of LID scenarios: Rain gardens (RG), permeable pavements (PP), and forested riparian buffers (RB). Our goal was to derive a relative understanding of how different spatial distributions of select LID types may affect hydrology in this mixed land over system; therefore, we did not aim to represent specific stakeholder-selected LID practices for the watershed (e.g., exact sites where landowners would agree to implementation).

To implement the LID scenarios, we replaced grid cells in the calibrated base model, identified as impervious, areas with one of two LID practices: RG or PP, depending on the impervious area type (see below). We further replaced grid cells in agricultural land cover along a stream with RB. We ran each scenario as a separate model using evenly distributed spatial configurations of 25%, 50%, 75% and 100% conversions for RG (in sidewalk locations) and PP (at parking lots and driveways; see Figure 3 for an example). Each spatial distribution of RG and PP met or exceeded the watershed's water quality volume for bioretention (i.e., generally speaking, the volume of water treated by LID practices to control in low to medium magnitude storm events), as recommended by the Ohio Department of Natural Resources [47]. We also placed RP at 20 m and 40 m on each side of the stream in the agricultural land of the Northern part of the watershed (Figure 1). This resulted in 10 simulated LID scenarios (4 RG, 4 PP, and 2 RB) for comparison. We note here that a large-scale conversion of impervious areas to LIDs (e.g., our 100% conversion scenarios) may not be reasonable, in terms of both financial cost and the willingness of the community [46]; however, these conversion configurations can provide a maximum mitigation potential for decision support.

RG and PP were chosen for the scenarios because they are reasonable retrofitting measures for the studied watershed, are the most promising LID practices for reductions in peak flow and runoff volume [8,48], and can be applied and assessed in the VELMA model. RBs were selected because they currently do not exist in the agricultural land of the watershed (and therefore the base model). Their addition was used for comparisons of the watershed-scale hydrological responses of RG and RB conversions on impermeable areas. We selected 20 m and 40 m buffers to go beyond Ohio EPA's requirement that forested area must be maintained for a minimum of the first 15 m of the area on either bank [49].

We assessed the effect of the LID scenario implementation: (a) Individually and (b) using an LID combination scenario (i.e., fully implementing RG, PP, and RB with the maximum level of implementation). The individual model scenario runs of land cover conversions to RG and PP, for each spatial configuration, included: (a) Sidewalks were converted to RG and (b) parking lots and driveways were converted to PP (Table 4 and Figure 3). Lawns were not converted to RG. The percentage of the watershed that was converted to RG, PP, and RB practices at different implementation levels is shown in Table 5.

Table 4. LID configurations in the model and conversion levels.

Type of Current land use	Type of Spatial Configuration Maps Under LID Scenarios	Type of LID Practices	Conversion Level
Sidewalks	Soil map, cover map, and permeability fraction map	Rain Garden (RG)	25%, 50%, 75%, 100%
Parking Lots and Driveways	Soil map and permeability fraction map	Permeable Pavement (PP)	25%, 50%, 75%, 100%
Agriculture	Soil map, cover map, and permeability fraction map	Riparian Buffer (RB)	20 m and 40 m

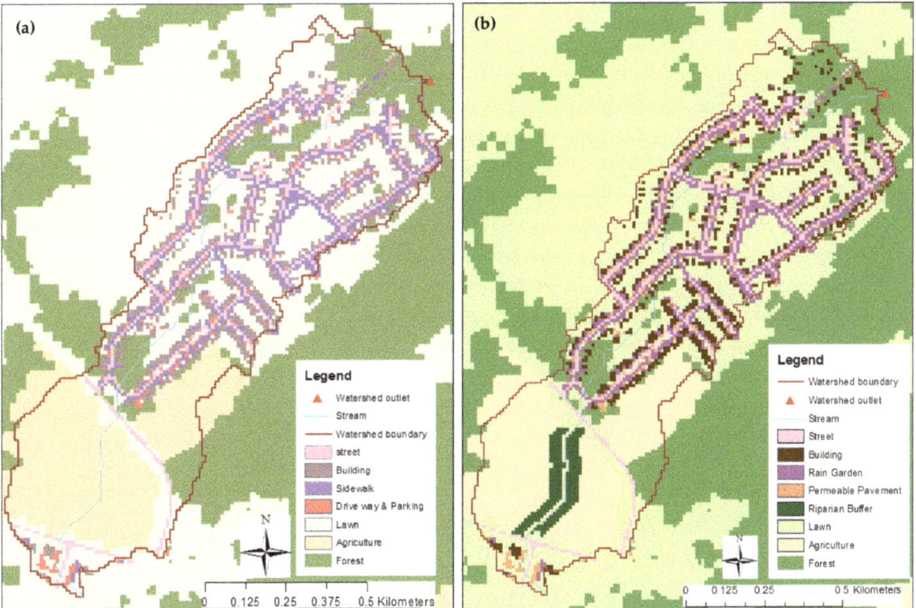

Figure 3. The spatial configuration of different land covers in the SHC watershed, with 10m cell resolution: (**a**) Current land cover, and (**b**) after LID implementation: Conversion of a 100% spatial configuration of the sidewalks to rain gardens, and driveways and parking lots into permeable pavements, and the implementation of 40 m forest buffers along both sides of the stream on the agricultural land. The watershed outlet is identified as a red triangle.

Table 5. The percent of the watershed converted to each LID practice and implementation level.

Type of LID Practices	Percent Watershed Converted			
	25%	50%	75%	100%
RG	1.6	3.2	4.8	6.4
PP	0.5	1.1	1.6	2.1
	20 m	40 m	-	-
RB	1.5	3.0	-	-

To implement LID into each scenario, we parameterized the soil texture, soil physical characteristics, and PET parameter values for LID practices, based on Ohio EPA requirements (Table 6) [49]. To do this for the RG scenarios, we created soil maps with a new soil class, "RG," for each spatial configuration (i.e., 25% to 100%). The RG soil maps, one for each implementation level, replaced the sidewalk pixels of the original soil map. Soils in the new RG maps were adjusted for soil depths, texture classes, and physical parameters to represent soils associated with rain gardens. The RG soil maps were based on Ohio EPA requirements for rain gardens (Table 6), which suggest that the soil media depths of a rain garden are 60–100 cm deep with loamy sand [49]. In the updated model configurations for each implementation scenario, we assumed no underdrain pipes and no outlet pipes, which are currently not implementable in VELMA.

In addition to the soil maps, we created new land cover maps for each spatial implementation level of RG (25% to 100%). We defined a new land cover, "RG," where existing sidewalks were located. For example, at the 50% RG implementation level, 50% of sidewalk's pixels of original were defined as "RG" land cover. For each new "RG" map, we parameterized the PET parameters of "RG" land cover to lawn values (Table 7).

For PP scenarios, we generated soil maps with a new soil class, "PP" which replaced parking lots and driveways at each conversion level (25% to 100%). We modified the original soil depths, soil texture classes, and soil physical parameter values for the "PP" soil class (Table 6) using the same values at each conversation level. According to the Ohio EPA Stormwater Management Practices manual, the recommended thickness of a PP system is 40–76 cm, depending on frost depth [49]. Therefore, for PP, we parameterized the hydraulic conductivity and other soil physical parameter values of the first 100 cm of the "PP" soil class [32]. We assumed that permeable pavement is a continuous pavement system (gravel) and well maintained with no clogging issues.

For RB, we created soil maps with a new soil class (Table 6) and a new land cover class (Table 7), "RB" to replace current soils and land cover at 20 m and 40 m on each side of streams where agriculture exists. Because most riparian buffers for Ohio streams are forested [49], we parameterized the soil parameters and PET values of the buffer area in the new soil and cover maps to reflect the effect of a forest rooting system and forest canopy on infiltration and ET (Tables 6 and 7).

For RG, PP and RB, new permeability fraction maps were also created for each implementation level to replace the original permeability fraction map in each model scenario. In the new permeability fraction maps, permeability fractions of 0 for impervious surfaces, such as sidewalks, parking lots and driveways, were changed to 1 and 0.95 for RG and PP, respectively. The permeability fraction for the RB scenario was changed from 0.95 for agriculture to 1 for RB forested land cover.

Once the base model was calibrated, we ran the model for each of the 10 scenarios under the different LID spatial configurations to evaluate changes in peak flows, surface runoff, ET, subsurface runoff and infiltration, and compared them to that of the base model (existing conditions).

Table 6. Soil parameter values used in the LID practice scenarios. RG: Rain Garden, PP: Permeable pavement, and RB: Riparian Buffer.

Parameter	Description	LID Practice	Layer	Value	Unit
z	Soil layer thickness [1]	All	1, 2, 3, 4	500, 500, 12,000, 12,000	mm
$K_{s,l}$	Saturated lateral hydraulic conductivity [2]	RG	1, 2, 3, 4	200, 200, 80, 30	-
		PP	1, 2, 3, 4	250, 250, 80, 30	
		RB	1, 2, 3, 4	300, 300, 80,30	
$K_{s,v}$	Saturated vertical hydraulic conductivity [2]	RG	1, 2, 3, 4	50, 50, 10, 10	-
		PP	1, 2, 3, 4	100, 100, 10, 10	
		RB		50, 30, 15, 10	
n	Porosity fraction [2]	RG	All	0.437	-
		PP		0.437	
		RB		0.437	
P_b	Bulk density [2]	RG	All	1.65	g cm^{-3}
		PP		1.65	
		RB		1.65	
θ_{wp}	Wilting point [2]	RG	All	0.055	-
		PP		0.033	
		RB		0.055	
θ_{fc}	Field capacity [2]	RG	All	0.125	-
		PP		0.091	
		RB		0.125	

[1] Ohio EPA [49]; [2] McKane et al. [32], based on the Ohio EPA recommended soil texture class of loamy sand and sand for RG and PP [49]. For the RB scenario, the values were set to the loamy sand soil texture class to represent a forest rooting system [32].

Table 7. Potential Evapotranspiration parameter values used in the LID practice scenarios. RG: Rain Garden, PP: Permeable Pavement, and RB: Riparian Buffer.

Parameter	Description	LID Practice	Value	Unit
PetParam1 [1]	First term of PET *Hamon* equation	RG	0.15	-
		PP	0.05	
		RB	0.30	
PetParam2	Second term of PET *Hamon* equation	RG		-
		PP	0.622	
		RB		
TemperaturePetOff	PET is only active when air temperature is greater than this value	RG		C°
		PP	−3	
		RB		
roair	Air density	RG		g m^{-3}
		PP	1300	
		RB		
be	ET coefficient used in the logistic equation that computes ET from PET	RG		-
		PP	3.07	
		RB		
noTranspirationPetFraction	The fraction of PET available outside of this cover's growing season	RG		-
		PP	1	
		RB		

[1] The values for RG and RB were set to the calibrated values for lawn and forest land cover (Table 3). The value for PP set to minimum value for impervious area (Table 3).

3. Results

3.1. Calibration and Validation of the Base Model

Daily streamflow calibration suggests acceptable model results across the simulation period (Figure 4a–c). The NSE, R^2, root mean square error (RMSE), and percent bias (PBIAS) for the calibration period (2009 and 2010) were 0.50, 0.53, 3.12 and −2.40, respectively. Moriasi et al. [50] recommended that an NSE ≥ 0.50 and PBIAS ≤ ±15 can be considered satisfactory for daily streamflow simulations. RMSE varies from 0 to large positive values. The lower the RMSE, the better the model fit [46]. The optimum value for PBIAS is zero, and low magnitude values indicate better simulations. Negative values indicate model overestimation [51]. While the daily model calibration is acceptable, it tends toward underestimating peak flows (Figure 4a,b). This is confirmed by a negative PBIAS; however, the magnitude is low, which means that the bias toward peak flow underestimation is minimal. Further, the Sim:Obs were 0.77, 1.10 and 0.96 (for 2009, 2010, 2011, respectively), all of which indicated that annual volumetric streamflow estimates in the base model were satisfactory.

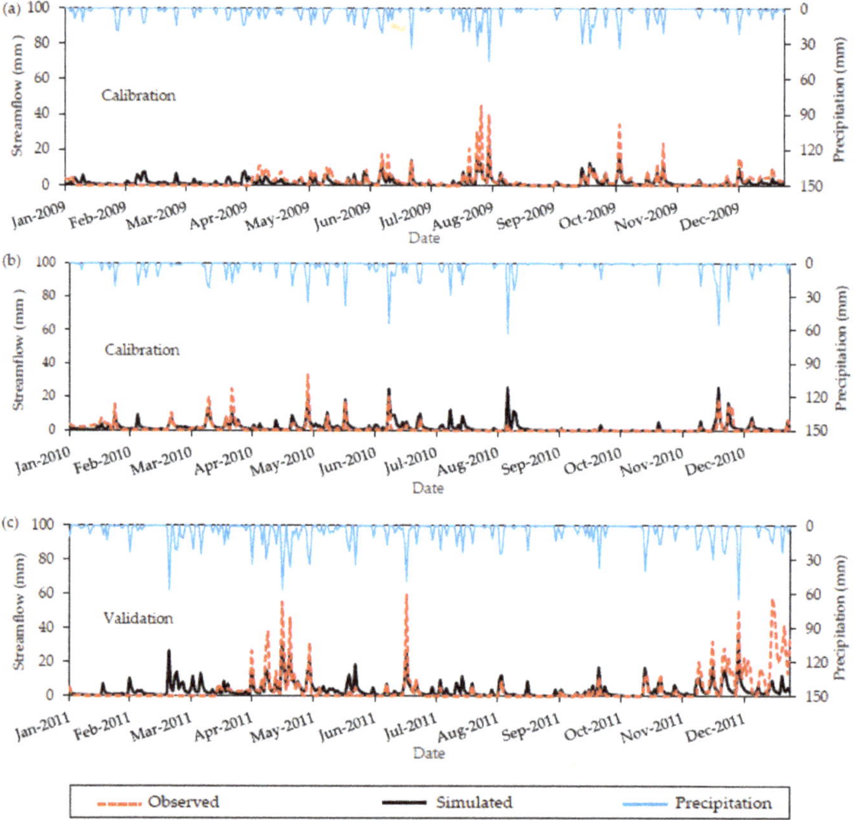

Figure 4. Plots of (**a**,**b**) calibration (NSE = 0.54, R^2 = 0.53, RMSE = 3.12, and PBIAS = −2.40) and (**c**) validation (NSE = 0.40, R^2 = 0.48, RMSE = 5.27, and PBIAS = 13.84) of the VELMA model output at the watershed outlet. The model was calibrated at a daily time step from 1 January 2009 to 31 December 2010 and validated at a daily time step from 1 January 2011 to 31 December 2011.

Simulations during the validation period captured general daily streamflow patterns; however, the model fit was less satisfactory than the calibration period (NSE of 0.40, R^2 of 0.48, RMSE of 5.27, and PBIAS of 13.84). Moreover, visual inspection of the validation plot (Figure 4c) indicated that the calibrated parameters were less successful during 2011, suggesting that calibrated model simulations may have increased limitations during wet years.

The average annual water balance components of the calibrated base model across the watershed, for the simulation period (2009–2011), are shown in Table 8. Evapotranspiration accounts for about 44% of precipitation, which approximates the lower end of the Sanford and Selnick [52] estimates of the fraction of precipitation lost to evapotranspiration in Southwest Ohio, USA.

Table 8. Average annual water balance components of the base model for the entire watershed, from 2009–2011. Note that precipitation is lower than the sum of the other water balance components because of the structure of VELMA's hydrological model output, which couples shallow subsurface runoff with infiltration.

Water Balance Component	Value (mm)
Precipitation	1249
Surface runoff	444
ET	548
Shallow subsurface runoff + infiltration	427

3.2. LID Scenarios

We compared the simulated water balances for the three LID practices at 25%, 50%, 75% and 100%, 20 m and 40 m implementation levels and one combined LID scenario at the maximum level of implementation. Our results suggest that LID practices decreased surface runoff and peak flow, and increased ET, shallow subsurface runoff and infiltration as the LID implementation level increased (Figure 5). However, the response varied among different LID practices (Figure 5).

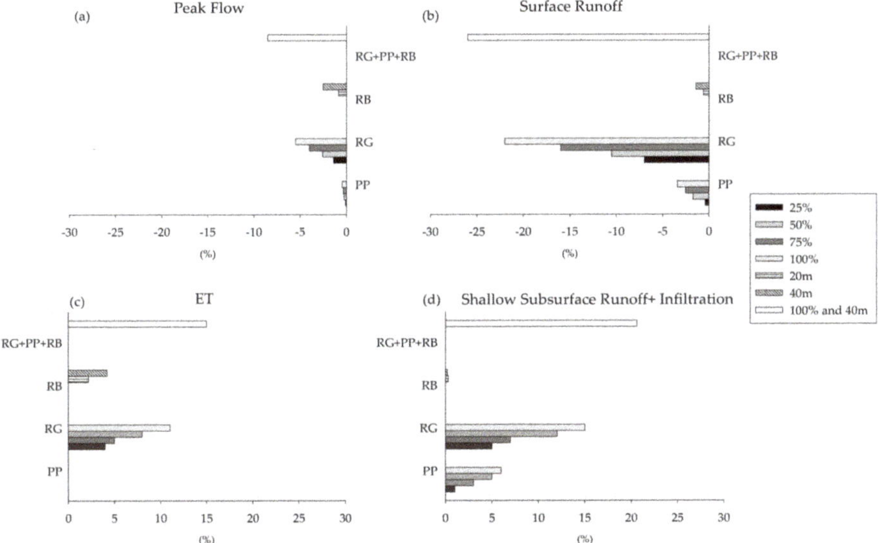

Figure 5. Percent change for watershed water balance components for three different types of LID practices at the outlet of the watershed (RG: Rain Garden, PP: Permeable Pavement, and RB: Riparian Buffer), (**a**) peak flow; (**b**) surface runoff; (**c**) ET (evapotranspiration); and (**d**) shallow subsurface runoff and infiltration. At the maximum level of implementation (100% and 40 m) RG, PP, and RB cover 6.4%, 2.1%, and 3% of the total watershed area, respectively.

Reduction in peak flows varies from about 0.5% to 5.5% among all individual LID practice scenarios, with the high reduction observed for the RG scenario at 100% and 75% implementation levels (5.5% and 4%, respectively), followed by 50% RG and 40 m RB scenarios (Figure 5a). Surface runoff decreased across all LID scenarios, with the largest reductions resulting from the RG scenario at 25%, 50%, 75% and 100% implementation levels (7%, 10.5%, 16% and 22%, respectively) (Figure 5b). PP and RB scenarios showed smaller reductions in surface runoff, ranging from 0.4 (for 25% PP) to 3.4% (for 100% PP). Reductions in surface runoff for 40 m and 20 m RB scenarios were 1.4% and 0.6%, respectively (Figure 5b). The percentage reduction in surface runoff was more than peak flows across all scenarios.

Retrofitting the baseline model with the LID increased shallow subsurface runoff and infiltration with increasing implementation levels as shown in Figure 5c,d. ET increased 2–15% for RG and RB scenarios across implementation levels, with higher increases in the 100%, 75% and 50% RG scenarios (11%, 8% and 5%, respectively). The RG scenario resulted in higher increases for both processes in comparison to other individual scenarios. Following the same trend, PP and RB scenarios increased shallow subsurface runoff and infiltration, ranging from 0.2% to 6% for different implementation levels (Figure 5d). Changes in ET for PP scenarios were negligible (Figure 5c).

Combining the three LID practices (RG, PP, and RB) at the highest implementation levels (100% for RG and PP, and 40 m for RB) resulted in the largest reductions in peak flows and surface runoff compared to individual LID implementations. The reductions in peak flow (8.5%) were modest, but considerably greater in surface runoff (26%; Figure 5a,b). The combined LID scenario resulted in the greatest increase in ET (15%), as well as a shallow subsurface runoff and infiltration (21%), in comparison with individual LID scenarios (Figure 5c,d).

The RG scenario showed the highest reduction in peak flows in comparison with PP and RB scenarios (Figure 5a). Therefore, we compared the peak flow to the percent of reduction in peak flow after RG implementation (100% scenario) during the simulation period (Figure 6). Peak flows were defined as one standard deviation above the mean simulated daily streamflow (here, 3.18 mm). We also considered the streamflow one day after we considered the peak flows to include a portion of the falling limb of the hydrograph. The percentage reduction in peak flows after RG implementation decreased exponentially with increasing peak flow conditions R^2 of 0.47; *p*-value < 0.001 (Figure 6).

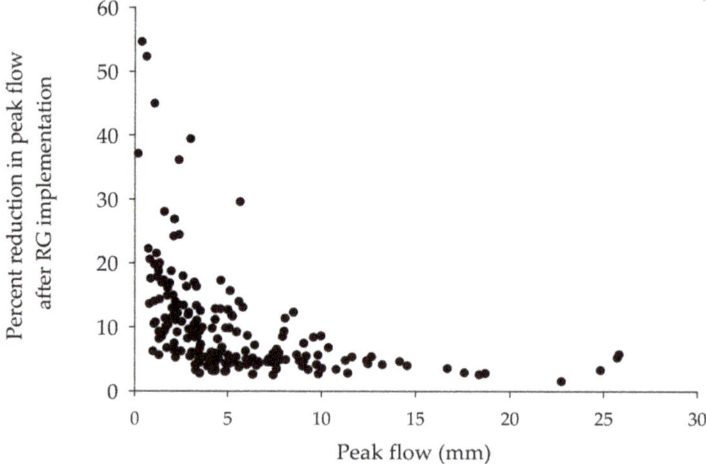

Figure 6. Percentage reduction in peak flow after RG implementation (100% scenario) vs. peak flow (mm) during the simulation period at the watershed outlet. Peak flow was estimated as any flow that was one standard deviation above the mean simulated daily flow for the study period or flow on the day following peak flow conditions, as described (to capture part of the falling limb of the hydrograph).

4. Discussion

4.1. LID Practices and Watershed-Scale Hydrological Effects

We assess, via spatially explicit model simulations, the relative effects of different types and configurations of LID practices on watershed hydrology in a mixed land cover system. Model simulation results suggest reductions in peak flows and surface runoff, and increases in evapotranspiration and subsurface flow and infiltration, with all spatial configurations of LID at the watershed scale. This is consistent with Gagrani et al. [19], Fry and Maxwell [53], and Avellaneda et al. [54], who reported similar effects on water balance components and peak flows after the placement of different LID practices in urban watersheds, with 42–55 percent impervious surfaces and drainage areas ranging from 0.2 km^2 to 12 km^2.

The magnitudes of simulated water balance responses to LID placement in our watershed study were lower than other studies in strictly urban watersheds (e.g., Fry and Maxwell [53] and Avellaneda et al. [54]) and more similar to a pilot study in a small suburban watershed (1.8 km^2) of Cincinnati, Ohio, where retrofitted rain gardens and rain barrels did not result in substantial runoff reductions [55]. The more limited response in our study watershed reflects, in part, the smaller extent of urban and suburban land cover compared to studies in other watersheds. Only 27 percent of our study watershed was covered in impervious surfaces, and only 31 percent of this area was converted to LID at the highest level of implementation. Therefore, our results are not completely unexpected and may point to important scale issues regarding the extent to which LID influences hydrologic regimes in mixed land use watersheds.

The simulated RB scenario did not result in a significant effect on peak flow at the watershed outlet and other water balance components. This is likely because only 3% of the total watershed area (and 13% of the watershed's agricultural land cover) was converted to RB at the highest implementation level (40 m). This indicates that the type and extent of LID practices affects cumulative watershed-scale hydrological effects [15,56].

In the mixed land cover SHC watershed, model comparisons among LID scenarios suggested that the RG was most effective across all implementation levels at reducing runoff and peak flow, and promoting ET, compared to PP and RB. RG also exerted greater control over modifying watershed water balance components, in terms of per unit area LID conversions. For example, at the 100% implementation level, RG reduced peak flows by 73% and 2%, decreased surface runoff by 50% and 86%, and increased ET by ~100%, which was 23% more than PP and RB, respectively. PP was 28% more effective at increasing shallow subsurface runoff and infiltration than RG. Recent studies point to a similar effectiveness of RG on water balance components at watershed scales [16,18]. Studies also have shown that PP can effectively mitigate surface runoff [57,58]; however, the degree of RG and PP functionality depends on the extent of the application area of LID within the watershed [59].

We found that 100% implementation of RG across the watershed was more effective at reducing peak flows during small storms than during larger ones (Figure 6). At the plot scale, Speak et al. [60] found that a green roof runoff retention significantly decreased during high rainfall events. These indicate that the effectiveness of any type of management practice, including LID, may be exponentially diminished as it loses storage capacity and becomes saturated. This counters Wadzuk et al. [61], who concluded, at the plot scale, that antecedent soil moisture conditions and "back-to-back events" are not a primary concern for biofiltration rain garden and green roof practices in recovering their infiltration capacity. However, based on our results, this finding may not be transferable to watershed scales, especially when LID practices receive both precipitation and appreciable surface runoff loading [62]. Our findings also highlight the potential importance of RG in controlling the first flush of pollutant loads and channel erosion [63,64] during more frequent storm events and a need for future research on the impact of LID practices across variable sequences of wet weather events.

4.2. Implications for Stormwater Management and Future Research

Our study provides scientists and watershed managers a glimpse into the potential influence of LID practices in mixed land cover systems, where only a portion of the watershed is converted to LID. Watershed-scale models, such as VELMA, can provide a physically-based and systematic means of projecting and evaluating the influence of various LID configurations in heterogeneous watersheds. Using this approach, our results suggest modest to minimal changes in most components of the water balance in response to LID, though these responses may be much more considerable if the watershed was exclusively urban or suburban land cover and converted to LID.

It is important to note that the location of the LID implementation with respect to the watershed outlet may also be critical [56]. For example, in our study, the watershed in agricultural areas are in the headwaters and LID implementation is in the lower portion of the watershed. Therefore, while our results suggest a limited shift in watershed hydrological dynamics with LID implementation, if LID was implemented into the upgradient of agricultural or forested land, then the magnitude of the response may be even less substantial due to attenuation from downgradient watershed processes. Based on these results, we suggest that future research needs to evaluate the hydrological effects of LID using distributed models, with a particular focus on how configurations of different land cover types influence watershed-scale LID responses, retiming of runoff delivery from subbasins of differing land cover, and antecedent soil moisture, as affected by storm sequences.

Model selection for assessing the hydrological effects of watershed-scale LID implementation is challenging because it involves trade-offs in achieving the necessary fidelity (i.e., the extent to which the model faithfully represents the modeled system) to hydrological, biological, and biogeochemical processes for prediction accuracy, while minimizing complexity and uncertainty [14]. Careful consideration of these tradeoffs is needed for future work that addresses how LID affects watershed-scale hydrological processes, particularly in mixed land cover systems. For example, existing models that explicitly integrate LID practices have been developed for urban systems and have specific LID modules for urban-based hydrological processes (e.g., SWMM and Green Infrastructure Flexible Model (GIFMod) [65]). On the other hand, models that have been explicitly developed to assess the effects of LID practices in mixed land cover watersheds (e.g., VELMA and Regional Hydro-Ecological Simulation System (RHESSys) [66]) may have a strong biogeochemical module (because of their mixed land cover focus) but a more limited hydrological capacity to physically represent LID practices, as compared with a model such as SWMM. Responses to these challenges are evolving by incorporating LID modules within ecohydrological models, such as VELMA, that provide mechanistic representations of LID performance [14] and coupling models to quantify how LID affects the fate and transport of various pollutants, as well as couple SWMM with other watershed models to improve simulations of urban hydrology [67].

Our findings suggest a clear need for the evaluation of the influence and benefits of LID in the context of other watershed land uses and their associated management [68]. For example, the incremental influence of LID on overall watershed responses, relative to management targets at different locations along the stream network, should be assessed. In this example, if a management goal is peak attenuation at the watershed outlet, a cost-benefit analysis, of how "best" to manage diffuse sources of runoff across different land cover types for peak flow reduction, would be beneficial.

Advancing the scientific understanding of the hydrological responses to LID in mixed land cover systems and linkages with the provision of diverse benefits is imperative because of the large number of watersheds globally that have mixed land cover. Future research may focus on upscaling fine scale studies to watersheds, applying a host of hydro-ecological models with LID modules or model parameter representations to address LID challenges in suburban watersheds (e.g., to provide multiple lines of evidence to support predicted outcomes), understanding LID's role in modifying baseflow (e.g., Bhaskar et al. [20]), and advancing these studies across diverse physiographic regions. Furthermore, given that we simulated and interpreted LID effects in this relatively small mixed land cover watershed, future research that applies continuous model simulations to project the hydrological effects of different LID configurations in mixed land cover watersheds with even greater complexity than ours will help to set realistic expectations for long term LID performance in these systems. Finally, research is also needed that expands our approaches to quantifying the effects of LID practices on nutrient and sediment loads, and links LID modules within watershed-scale ecohydrology models with simulated or measured in-stream processes.

5. Conclusions

We provide one of the first studies, to our knowledge, that assesses the relative watershed-scale hydrological effects of different types and configurations of LID practices in a mixed land cover watershed using a spatially explicit modeling approach. We simulated 10 scenarios across multiple spatial configurations of LID to evaluate the watershed hydrological responses of three practices—rain gardens (RG), permeable pavements (PP), and riparian buffers (RB)—in a 0.92 km^2 watershed with mixed suburban, agricultural, and forest land cover. A spatially-explicit ecohydrological model (VELMA) was used to compare changes in the watershed's water balance before and after LID practice implementation.

Overall, we found that the type and extent of LID practices influence watershed hydrological responses in our study system. Our simulation results indicate that LID practices decreased surface runoff and peak flow, and promoted ET, shallow subsurface runoff and infiltration. However, hydrologic responses and effectiveness varied among LID practices and implementation levels. When LID practices were considered individually, on a LID per unit area basis across all LID implementation levels, RG was more effective in reducing runoff and peak flow, and promoting ET, than PP and RB. However, our results indicated that the 100% implementation of RG was more effective at reducing peak flows during small storms than larger ones, suggesting that LID storage capacities are reduced due to soil saturation during and following large events. Further, both RG and PP increased shallow subsurface runoff and infiltration to almost the same extent at the watershed outlet and the combined LID scenario resulted in the highest performance by increasing shallow subsurface runoff and infiltration, and evapotranspiration by 21% and 15%, respectively, and reductions in peak flow and surface runoff of 8.5% and 8%, respectively.

We conclude that the spatial configurations and extent of LID practices, as well as the model selection and degree of watershed heterogeneity, might be critical for assessing the hydrological responses of watershed-scale LID implementation and must be considered in future research. Further research is needed to apply different LID configurations within mixed land cover watersheds to better understand LID performance and to evaluate the effect of LID practices on nutrient and sediment loads.

Author Contributions: H.E.G., C.T.N. and N.H. designed the project and modeling approach; N.H. performed, and R.B.M., B.L.B., A.F.B., K.S.D., J.J.H. and P.P.P. assisted with, the model set up, simulation, and calibration; N.H., H.E.G. and B.P.B. analyzed the results and developed the manuscript's structure; all authors contributed to writing the paper.

Acknowledgments: The authors would like to thank Amr Safwat with the CBI Federal Services and Amy Prues with Pegasus, Inc., in Cincinnati, Ohio, and Joong Lee in the Center for Urban Green Infrastructure Engineering, Milford, Ohio, for providing base maps, LID practice configurations in ArcGIS, and impervious area maps.

References

1. Shuster, W.; Bonta, J.; Thurston, H.; Warnemuende, E.; Smith, D. Impacts of impervious surface on watershed hydrology: A review. *Urban Water J.* **2005**, *2*, 263–275. [CrossRef]
2. Booth, D.B.; Jackson, C.R. Urbanization of aquatic systems: Degradation thresholds, stormwater detection, and the limits of mitigation. *J. Am. Water Resour. Assoc.* **1997**, *33*, 1077–1090. [CrossRef]

3. Zhang, K.; Chui, T.F.M. A comprehensive review of spatial allocation of lid-bmp-gi practices: Strategies and optimization tools. *Sci. Total Environ.* **2018**, *621*, 915–929. [CrossRef] [PubMed]

4. Fletcher, T.; Andrieu, H.; Hamel, P. Understanding, management and modelling of urban hydrology and its consequences for receiving waters: A state of the art. *Adv. Water Resour.* **2013**, *51*, 261–279. [CrossRef]

5. Fletcher, T.D.; Shuster, W.; Hunt, W.F.; Ashley, R.; Butler, D.; Arthur, S.; Trowsdale, S.; Barraud, S.; Semadeni-Davies, A.; Bertrand-Krajewski, J.-L. SUDS, LID, BMPs, WSUD and more–the evolution and application of terminology surrounding urban drainage. *Urban Water J.* **2015**, *12*, 525–542. [CrossRef]

6. Miles, B.; Band, L.E. Green infrastructure stormwater management at the watershed scale: Urban variable source area and watershed capacitance. *Hydrol. Processes* **2015**, *29*, 2268–2274. [CrossRef]

7. Collins, K.A.; Lawrence, T.J.; Stander, E.K.; Jontos, R.J.; Kaushal, S.S.; Newcomer, T.A.; Grimm, N.B.; Ekberg, M.L.C. Opportunities and challenges for managing nitrogen in urban stormwater: A review and synthesis. *Ecol. Eng.* **2010**, *36*, 1507–1519. [CrossRef]

8. Ahiablame, L.M.; Engel, B.A.; Chaubey, I. Effectiveness of low impact development practices: Literature review and suggestions for future research. *Water Air Soil Pollut.* **2012**, *223*, 4253–4273. [CrossRef]

9. Hamel, P.; Daly, E.; Fletcher, T.D. Source-control stormwater management for mitigating the impacts of urbanisation on baseflow: A review. *J. Hydrol.* **2013**, *485*, 201–211. [CrossRef]

10. Elliott, A.; Trowsdale, S. A review of models for low impact urban stormwater drainage. *Environ. Modell. Softw.* **2007**, *22*, 394–405. [CrossRef]

11. Jayasooriya, V.; Ng, A.W.M. Tools for modeling of stormwater management and economics of green infrastructure practices: A review. *Water Air Soil Pollut.* **2014**, *225*, 2055. [CrossRef]

12. Vogel, J.R.; Moore, T.L.; Coffman, R.R.; Rodie, S.N.; Hutchinson, S.L.; McDonough, K.R.; McLemore, A.J.; McMaine, J.T. Critical review of technical questions facing low impact development and green infrastructure: A perspective from the great plains. *Water Environ. Res.* **2015**, *87*, 849–862. [CrossRef] [PubMed]

13. Burns, M.J.; Fletcher, T.D.; Walsh, C.J.; Ladson, A.R.; Hatt, B.E. Hydrologic shortcomings of conventional urban stormwater management and opportunities for reform. *Landsc. Urban Plann.* **2012**, *105*, 230–240. [CrossRef]

14. Golden, H.E.; Hoghooghi, N. Green infrastructure and its catchment-scale effects: An emerging science. *Wiley Interdiscip. Rev. Water* **2018**, *5*, e1254. [CrossRef] [PubMed]

15. Bell, C.D.; McMillan, S.K.; Clinton, S.M.; Jefferson, A.J. Hydrologic response to stormwater control measures in urban watersheds. *J. Hydrol.* **2016**, *541*, 1488–1500. [CrossRef]

16. Jarden, K.M.; Jefferson, A.J.; Grieser, J.M. Assessing the effects of catchment-scale urban green infrastructure retrofits on hydrograph characteristics. *Hydrol. Processes* **2015**, *30*, 1536–1550. [CrossRef]

17. Yang, B.; Li, S. Green infrastructure design for stormwater runoff and water quality: Empirical evidence from large watershed-scale community developments. *Water* **2013**, *5*, 2038–2057. [CrossRef]

18. Pennino, M.J.; McDonald, R.I.; Jaffe, P.R. Watershed-scale impacts of stormwater green infrastructure on hydrology, nutrient fluxes, and combined sewer overflows in the mid-Atlantic region. *Sci. Total Environ.* **2016**, *565*, 1044–1053. [CrossRef] [PubMed]

19. Gagrani, V.; Diemer, J.A.; Karl, J.J.; Allan, C.J. Assessing the hydrologic and water quality benefits of a network of stormwater control measures in a se us piedmont watershed. *J. Am. Water Resour. Assoc.* **2014**, *50*, 128–142. [CrossRef]

20. Bhaskar, A.S.; Hogan, D.M.; Archfield, S.A. Urban base flow with low impact development. *Hydrol. Processes* **2016**, *30*, 3156–3171. [CrossRef]

21. Abdelnour, A.; Stieglitz, M.; Pan, F.; McKane, R. Catchment hydrological responses to forest harvest amount and spatial pattern. *Water Resour. Res.* **2011**, *47*. [CrossRef]

22. Beven, K.J. *Rainfall-Run off Modelling: The Primer*; John Wiley & Sons: New York, NY, USA, 2011.

23. US Environmental Protection Agency. Storm Water Management Model (SWMM). Available online: https://www.epa.gov/water-research/storm-water-management-model-swmm (accessed on 10 May 2018).

24. Arnold, J.G.; Srinivasan, R.; Muttiah, R.S.; Williams, J.R. Large area hydrologic modeling and assessment part I Model development. *J. Am. Water Resour. Assoc.* **1998**, *34*, 73–89. [CrossRef]

25. Seo, M.; Jaber, F.; Srinivasan, R.; Jeong, J. Evaluating the impact of Low Impact Development (LID) practices on water quantity and quality under different development designs using swat. *Water* **2017**, *9*, 193. [CrossRef]

26. Cipolla, S.S.; Maglionico, M.; Stojkov, I. A long-term hydrological modelling of an extensive green roof by means of SWMM. *Ecol. Eng.* **2016**, *95*, 876–887. [CrossRef]

27. Khader, O.; Montalto, F.A. In Development and calibration of a high resolution SWMM model for simulating the effects of LID retrofits on the outflow hydrograph of a dense urban watershed. In Proceedings of the 2008 International Low Impact Development Conference, Seattle, WA, USA, 16–19 November 2008.

28. Walsh, T.C.; Pomeroy, C.A.; Burian, S.J. Hydrologic modeling analysis of a passive, residential rainwater harvesting program in an urbanized, semi-arid watershed. *J. Hydrol.* **2014**, *508*, 240–253. [CrossRef]

29. Wong, T.H.; Fletcher, T.D.; Duncan, H.P.; Jenkins, G.A. Modelling urban stormwater treatment—A unified approach. *Ecol. Eng.* **2006**, *27*, 58–70. [CrossRef]

30. Lee, J.G.; Nietch, C.T.; Panguluri, S. Drainage area characterization for evaluating green infrastructure using the storm water management model. *Hydrol. Earth Syst. Sci.* **2018**, *22*, 2615–2635. [CrossRef]

31. Reyes, B.; Maxwell, R.M.; Hogue, T.S. Impact of lateral flow and spatial scaling on the simulation of semi-arid urban land surfaces in an integrated hydrologic and land surface model. *Hydrol. Processes* **2015**, *22*, 2615–2635. [CrossRef]

32. McKane, R.; Brookes, A.; Djang, K.; Stieglitz, M.; Abdelnour, A.; Pan, F.; Halama, J.; Pettus, P.; Phillips, D. *Velma User Manual and Technical Documentation*, 2nd ed.; U.S. Environmental Protection Agency Office of Research and Development National Health and Environmental Effects Research Laboratory: Corvallis, OR, USA, 2014.

33. Brockman, C.S. Physiographic Regions of Ohio. 1998. Available online: http://www.epa.state.oh.us/portals/27/SIP/SO2/D2_physio.pdf (accessed on 12 January 2017).

34. Braun, E.L. Forests of the Illinoian till plain of southwestern Ohio. *Ecol. Monogr.* **1936**, *6*, 89–149. [CrossRef]

35. United States Department of Agriculture, Natural Resources Conservation Service. Available online: https://websoilsurvey.sc.egov.usda.gov/App/WebSoilSurvey.aspx (accessed on 25 September 2016).

36. National Climatic Data Center. Available online: https://www.ncdc.noaa.gov/cdo-web/datatools/findstation (accessed on 12 January 2017).

37. Texas A&M University. Global Weather Data for Swat. Available online: https://globalweather.tamu.edu/ (accessed on 28 October 2016).

38. United States Department of Agriculture, Natural Resources Conservation Service. Available online: https://datagateway.nrcs.usda.gov/GDGOrder.aspx (accessed on 25 September 2016).

39. Hamon, W.R. Computation of direct runoff amounts from storm rainfall. *Int. Assoc. Sci. Hydrol. Publ.* **1963**, *63*, 52–62.

40. Abdelnour, A.; McKane, R.B.; Stieglitz, M.; Pan, F.; Cheng, Y. Effects of harvest on carbon and nitrogen dynamics in a pacific northwest forest catchment. *Water Resour. Res.* **2013**, *49*, 1292–1313. [CrossRef]

41. Pan, F.; Stieglitz, M.; McKane, R.B. An algorithm for treating flat areas and depressions in digital elevation models using linear interpolation. *Water Resour. Res.* **2012**, *48*. [CrossRef]

42. Framework. Multiobjective Evolutionary Algorithms. Available online: http://moeaframework.org/ (accessed on 14 November 2016).

43. Deb, K.; Pratap, A.; Agarwal, S.; Meyarivan, T. A fast and elitist multiobjective genetic algorithm: Nsga-II. *IEEE Trans. Evol. Comput.* **2002**, *6*, 182–197. [CrossRef]

44. Nash, J.E.; Sutcliffe, J.V. River flow forecasting through conceptual models part I—A discussion of principles. *J. Hydrol.* **1970**, *10*, 282–290. [CrossRef]

45. Gupta, H.V.; Sorooshian, S.; Yapo, P.O. Status of automatic calibration for hydrologic models: Comparison with multilevel expert calibration. *J. Hydrol. Eng.* **1999**, *4*, 135–143. [CrossRef]

46. Moriasi, D.N.; Arnold, J.G.; Van Liew, M.W.; Bingner, R.L.; Harmel, R.D.; Veith, T.L. Model evaluation guidelines for systematic quantification of accuracy in watershed simulations. *Trans. ASABE* **2007**, *50*, 885–900. [CrossRef]

47. Ohio Department of Natural Resources. *Rainwater and Land Development and Urban Stream Protection*, 3rd ed.; Division of Soil Conservation: Columbus, OH, USA, 2006. Available online: http://soilandwater.ohiodnr.gov/portals/soilwater/pdf/stormwater/Intro_3-3-14.pdf (accessed on 5 June 2018).

48. Dietz, M.E. Low impact development practices: A review of current research and recommendations for future directions. *Water Air Soil Pollut.* **2007**, *186*, 351–363. [CrossRef]

49. Ohio EPA. Ohio Environmental Protection Agency, Post-Construction Storm Water Practices. Available online: http://epa.ohio.gov/dsw/storm/technical_guidance.aspx#176135061-rainwater-and-land-development-manual (accessed on 14 November 2016).

50. Moriasi, D.N.; Gitau, M.W.; Pai, N.; Daggupati, P. Hydrologic and water quality models: Performance measures and evaluation criteria. *Trans. ASABE* **2015**, *58*, 1763–1785.

51. Gupta, H.V.; Kling, H.; Yilmaz, K.K.; Martinez, G.F. Decomposition of the mean squared error and NSE performance criteria: Implications for improving hydrological modelling. *J. Hydrol.* **2009**, *377*, 80–91. [CrossRef]

52. Sanford, W.E.; Selnick, D.L. Estimation of evapotranspiration across the conterminous united states using a regression with climate and land-cover data 1. *J. Am. Water Resour. Assoc.* **2013**, *49*, 217–230. [CrossRef]

53. Fry, T.J.; Maxwell, R.M. Evaluation of distributed bmps in an urban watershed-high resolution modeling for stormwater management. *Hydrol. Processes* **2017**, *31*, 2700–2712. [CrossRef]

54. Avellaneda, P.M.; Jefferson, A.J.; Grieser, J.M.; Bush, S.A. Simulation of the cumulative hydrological response to green infrastructure. *Water Resour. Res.* **2017**, *53*, 3087–3101. [CrossRef]

55. Shuster, W.; Rhea, L. Catchment-scale hydrologic implications of parcel-level stormwater management (Ohio USA). *J. Hydrol.* **2013**, *485*, 177–187. [CrossRef]

56. Di Vittorio, D.; Ahiablame, L. Spatial translation and scaling up of low impact development designs in an urban watershed. *J. Water Manag. Model.* **2015**, 1–9. [CrossRef]

57. Bedan, E.S.; Clausen, J.C. Stormwater runoff quality and quantity from traditional and low impact development watersheds1. *Jawra J. Am. Water Resour. Assoc.* **2009**, *45*, 998–1008. [CrossRef]

58. Ahiablame, L.M.; Engel, B.A.; Chaubey, I. Effectiveness of low impact development practices in two urbanized watersheds: Retrofitting with rain barrel/cistern and porous pavement. *J. Environ. Manag.* **2013**, *119*, 151–161. [CrossRef] [PubMed]

59. Her, Y.; Jeong, J.; Arnold, J.; Gosselink, L.; Glick, R.; Jaber, F. A new framework for modeling decentralized low impact developments using soil and water assessment tool. *Environ. Model. Softw.* **2017**, *96*, 305–322. [CrossRef]

60. Speak, A.; Rothwell, J.; Lindley, S.; Smith, C. Rainwater runoff retention on an aged intensive green roof. *Sci. Total Environ.* **2013**, *461*, 28–38. [CrossRef] [PubMed]

61. Wadzuk, B.M.; Lewellyn, C.; Lee, R.; Traver, R.G. Green infrastructure recovery: Analysis of the influence of back-to-back rainfall events. *J. Sustain. Water Built Environ.* **2017**, *3*, 04017001. [CrossRef]

62. Winston, R.J.; Dorsey, J.D.; Smolek, A.P.; Hunt, W.F. Hydrologic performance of four permeable pavement systems constructed over low-permeability soils in northeast Ohio. *J. Hydrol. Eng.* **2018**, *23*, 04018007. [CrossRef]

63. Davis, A.P.; Hunt, W.F.; Traver, R.G.; Clar, M. Bioretention technology: Overview of current practice and future needs. *J. Environ. Eng.* **2009**, *135*, 109–117. [CrossRef]

64. Hawley, R.J.; Bledsoe, B.P. How do flow peaks and durations change in suburbanizing semi-arid watersheds? A southern California case study. *J. Hydrol.* **2011**, *405*, 69–82. [CrossRef]

65. US Environmental Protection Agency. National Stormwater Calculator. Available online: https://www.epa.gov/water-research/national-stormwater-calculator (accessed on 10 May 2018).

66. University of North Carolina and San Diego State University. The Regional Hydro-Ecologic Simulation System (RHESSys). Available online: http://fiesta.bren.ucsb.edu/~rhessys/ (accessed on 10 May 2018).

67. Niazi, M.; Nietch, C.; Maghrebi, M.; Jackson, N.; Bennett, B.R.; Tryby, M.; Massoudieh, A. Storm water management model: Performance review and gap analysis. *J. Sustain. Water Built Environ.* **2017**, *3*, 04017002. [CrossRef]

68. Martin, K.L.; Hwang, T.; Vose, J.M.; Coulston, J.W.; Wear, D.N.; Miles, B.; Band, L.E. Watershed impacts of climate and land use changes depend on magnitude and land use context. *Ecohydrology* **2017**, *10*, e1870. [CrossRef]

Article

Modeling Landscape Change Effects on Stream Temperature Using the Soil and Water Assessment Tool

Mamoon Mustafa [1,*]**, Brad Barnhart** [2]**, Meghna Babbar-Sebens** [1] **and Darren Ficklin** [3]

[1] School of Civil & Construction Engineering, Oregon State University, Corvallis, OR 97330, USA; meghna@oregonstate.edu

[2] Western Ecology Division, National Health and Ecological Effects Laboratory, Office of Research and Development, U.S. Environmental Protection Agency, Corvallis, OR 97330, USA; bradleybarnhart@gmail.com

[3] Department of Geography, Indiana University, Bloomington, IN 47405, USA; dficklin@indiana.edu

* Correspondence: mustafamamoon90@gmail.com

Received: 13 July 2018; Accepted: 15 August 2018; Published: 27 August 2018

Abstract: Stream temperature is one of the most important factors for regulating fish behavior and habitat. Therefore, models that seek to characterize stream temperatures, and predict their changes due to landscape and climatic changes, are extremely important. In this study, we extend a mechanistic stream temperature model within the Soil and Water Assessment Tool (SWAT) by explicitly incorporating radiative flux components to more realistically account for radiative heat exchange. The extended stream temperature model is particularly useful for simulating the impacts of landscape and land use change on stream temperatures using SWAT. The extended model is tested for the Marys River, a western tributary of the Willamette River in Oregon. The results are compared with observed stream temperatures, as well as previous model estimates (without radiative components), for different spatial locations within the Marys River watershed. The results show that the radiative stream temperature model is able to simulate increased stream temperatures in agricultural sub-basins compared with forested sub-basins, reflecting observed data. However, the effect is overestimated, and more noise is generated in the radiative model due to the inclusion of highly variable radiative forcing components. The model works at a daily time step, and further research should investigate modeling at hourly timesteps to further improve the temporal resolution of the model. In addition, other watersheds should be tested to improve and validate the model in different climates, landscapes, and land use regimes.

Keywords: stream temperature; SWAT; Marys River watershed

---

1. Introduction

Stream temperature is an important water quality parameter that affects physical and chemical processes in streams [1]. Higher stream temperatures in river systems represent a growing concern worldwide and can affect the habitat and life spans of fish [2,3]. According to Eaton and Scheller [4], some fish species will disappear from the water body, if stream temperature transcends an upper limit. Using historical data ranging from 30 to 100 years, Kaushal et al. [5] reported that stream temperatures have been increasing throughout the United States at a rate of 0.009–0.077 °C/year, with a significant increase in the western United States. In particular, stream temperatures in the Pacific Northwest have reached historical records—at times, they have exceeded the lethal limit of 21.1 °C for some aquatic species such as salmon. For example, in the summer of 2015, the river temperature in the Columbia River in the State of Washington reached the level of 24.5 °C and led to the death of 235,000 sockeye salmon out of the total 507,000 that passed through the Bonneville Dam [6].

Similarly, Marys River (Hydrologic Unit Code 17090003) is a tributary of the Willamette River in Northwest Oregon and has experienced increasing temperatures over the last few years. The Oregon Department of Fish and Wildlife (ODFW) conducted a stream survey on flow conditions between the years of 1991 and 1993 and observed that the maximum stream temperature reached the maximum limit of 17.8 °C (ODFW). However, recently, increasing levels of human activities have resulted in even higher water temperatures. For example, high water temperatures between 21.1 °C and 26.7 °C were observed from June to August for some tributaries of Marys River (Marys River Watershed Council), which has resulted in several United States Environmental Protection Agency (USEPA) 303(d) listings for temperature exceedances. These trends may be related to changing climatic drivers as well as land use practices (e.g., harvest of timber, increasing barren lands and clear-cutting areas throughout the watershed) and landscape changes (e.g., urban and agricultural development).

Amidst large-scale landscape and land use changes, preservation of riparian buffers can increase stream shading, thereby helping regulate water temperature along stream reaches [7]. Stream shading intercepts and absorbs a large portion of solar radiation before it reaches the water surface, resulting in less thermal energy that reaches and is stored in streams, which indirectly helps to cool stream temperatures. Brown et al. [8] conducted a study in the Alsea watershed in Oregon along the coast range to study the impact of shading on stream temperatures before and after clear cuts in the watershed. They found that clear-cutting resulted in stream temperature increases of 7.8 °C one year after the cuts. Bond et al. [9] investigated the impact of riparian reforestation on summer stream temperatures in the Salmon River in northern California, and they found that partial reforestation lowered stream temperatures by 0.11–0.12 °C/km and by 0.26–0.27 °C/km for full reforestation.

Since increasing water temperatures have remained a major concern in many watersheds, many models have been proposed to simulate stream temperatures at time scales varying from minutes to months (see Ficklin et al. [10] for a brief review). For this paper, we specifically focus on a semi-distributed mechanistic watershed model called the Soil and Water Assessment Tool (SWAT) [11], which has been extensively used to evaluate the effects of landscape and land use changes on different hydrologic components. Ficklin et al. [10] improved the original stream temperature model within SWAT, which was a linear regression model by Stefan and Preud'homme [12] that correlates thermal energy exchange of air temperature to water temperature. Ficklin et al. [10] developed a daily-scale model for stream temperature prediction by integrating multiple climate and hydrological components, including snowmelt, surface runoff, lateral flow, groundwater flow, and finally air temperature. Several studies have found the Ficklin et al. model [10] produces more realistic simulation results compared to the linear regression proposed by Stefan and Preud'homme [12] (Barnhart et al., 2014; Ficklin et al., 2012; Ficklin et al., 2014 [10,13,14]). However, the model developed by Ficklin et al. [10] does not explicitly account for the different types of radiation that affect thermal energy of water systems, and only recent work has attempted to improve the model by incorporating select radiative components [15]. In general, thermal energy added and removed from any water system consists of incoming radiation that adds thermal energy to the water and results in increasing stream temperatures. This incoming radiation mainly consists of solar radiation coming from the sun, atmospheric longwave radiation, landcover longwave radiation, convection, and evaporation. In contrast, backscattered radiation removes thermal energy and helps to cool temperatures. This radiation consists of emitted longwave radiation from the water surface as well as convection and evaporation. Convection and evaporation radiative components can either add or remove energy from any water body depending on stream temperatures and climate conditions, specifically air temperature, humidity, and wind speed.

This paper highlights model development to explicitly incorporate multiple thermal radiation components into the Ficklin et al. [10] stream temperature model within SWAT. These thermal radiation equations are used within the widely used HEATSOURCE model [16], but until now, these equations have yet to be incorporated into SWAT. HEATSOURCE differs from SWAT because it is a reach-based stream temperature model, whereas SWAT is a watershed model. This means that SWAT simulates upland processes in addition to in-stream processes. HEATSOURCE can model stream temperatures at hourly timesteps and requires site-specific data (e.g., shading, canopy structure, stream morphology) that is oftentimes not available over the entire spatial extent of watersheds. Conversely, SWAT simulates hydrologic components and stream temperatures throughout a watershed using a daily time step and utilizes generally obtainable input data, such as spatially distributed precipitation, temperature, elevation, land use, and soil type. Users may prefer to use SWAT instead of HEATSOURCE when site-specific data is unavailable or when the study goal is to determine the effect of alternative land management scenarios on stream temperatures throughout large, heterogeneous watersheds.

The paper is organized as follows. First, the study area and the SWAT model setup are described. Then, three different SWAT stream temperature models are analyzed, including Stefan and Preud'homm [12] air temperature regression, a mechanistic model by Ficklin et al. [10], and an extension of the Ficklin et al. [10] model in which we specifically incorporate radiative components. We calibrate SWAT for hydrologic discharge in the Marys River watershed, and we compare the three stream temperature models to examine their relative performance for multiple sub-basins with different land use/land cover. We demonstrate the utility of our results by comparing simulations for sub-basins within primarily forested and agricultural landscapes.

2. Methodology

2.1. Study Area

The Marys River watershed, shown in Figure 1, is located in the Pacific Northwest of the United States (Hydrologic Unit Code (HUC) 17090003) and is part of the Willamette River basin (HUC 170900) in Oregon. It is one of five major river systems located on the western side of the Willamette River. The area of the watershed is of 782 square kilometers. The highest point of the watershed is at Marys Peak at an elevation of 1280 meters above sea level, and the lowest point is in Corvallis, Oregon, where Marys River drains into the Willamette River at an elevation of 76 meters above sea level. The climate of the watershed in the winter season is mild and wet, with an average winter temperature of 5 °C and rainfall during the winter ranges from 1000 mm downstream of the watershed to more than 2500 mm at the highest elevation upstream of the watershed. In general, the watershed tends to be dry, sunny, and warm throughout the summer (Marys River Watershed Council). It has an average summer temperature of 17.5 °C. High rainfall intensity results in high stream discharge during winter and spring and mean annual flows of 12–13 m^3/s. However, during the summer, flows are generally very low, and discharge sometimes drops below one cubic meter per second. Base flow is a major contributor to the flow of the river, where 61–70% of the total stream flow comes from groundwater contributions [17].

The watershed is divided into three different land use categories: Forest, agricultural, and urban. Most of the watershed (65% of the total area) consists of forest, which is largely located along the western portion of the watershed. In these mixtures of deciduous and evergreen forests, small streams flow over beds of gravel and cobbles with high velocities due to steep slope gradients. Flow leaving the forested region then enters agricultural land in the Willamette Valley that consists mainly of cultivated crops, hay, pasture, wheat, and grass seed production. The streams in this region flow on sand and silt with mild slope gradients, resulting in decreased flow velocities. Furthermore, urban areas are situated further downstream within the Willamette Valley (e.g., the cities of Philomath and Corvallis), and the stream flows over mostly flat to nearly flat gradients. Stream velocities decrease significantly as Marys River enters Philomath, Oregon, and then continues eastwards into Corvallis, Oregon, until it meets the Willamette River at the lowest point located in the watershed.

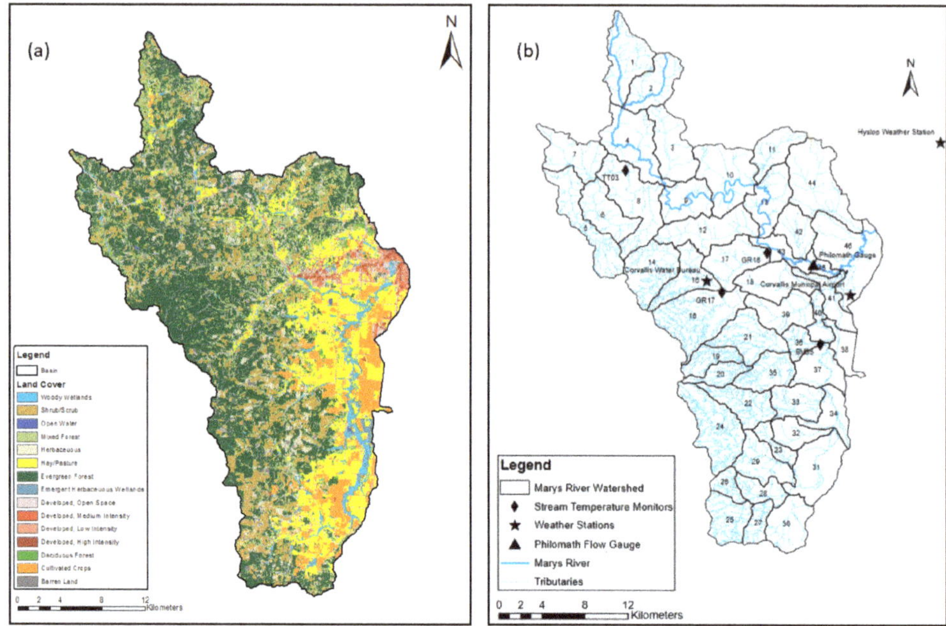

Figure 1. (**a**) Overview of Marys river watershed, and (**b**) land use in Oregon, USA.

2.2. Soil and Water Assessment Tool (SWAT)

The Soil and Water Assessment Tool (SWAT) is a semi-distributed watershed model that is designed to predict the impact of management on water, sediment, and agricultural chemical yields in gauged and ungauged watersheds [11]. In this study, SWAT was used to simulate the hydrologic and stream temperature dynamics within the Marys River watershed and to evaluate a model extension to the stream temperature model developed by Ficklin et al. [10]. The Marys River watershed was divided into smaller sub-watersheds using pre-defined drainage boundaries and a 10-meter digital elevation model, and then the sub-watersheds in the SWAT model were further divided into smaller units called hydrologic response units (HRUs) using ArcSWAT, a toolbox within ArcGIS for SWAT, with a HRU percentage threshold of 5%. Each HRU is a unique combination of land-use, soil type, and topographic slope and represents the basic unit for conducting mass balances and hydrologic flow in SWAT. The area of the 46 predefined sub-basins varies from 35 square kilometers for the largest sub-basin to 3.0 square kilometers for the smallest. The average sub-basin area is 17 square kilometers. The watershed slope was divided into two categories: (1) A steep gradient area located mostly within the forested regions in the western portion of the watershed, and (2) the nearly flat region located within the Willamette valley, east of the watershed where the cities Philomath and Corvallis are located.

SWAT's input data types include spatial GIS input files such as a Digital Elevation Model (DEM), a land use land cover layer, and a soil layer [18]. Input data needed to delineate the watershed including the DEM, sub-basins, and stream layers in addition to necessary land use and soil SSURGO (Soil Survey Geographic Database) layers to build the HRUs were acquired from United States Department of Agriculture [19]. Three weather stations were used as climate forcings: Corvallis Water Bureau (CWB) COOP ID of (351877), Hyslop weather station, which is also known as Oregon State University weather station (OSU) COOP ID of (351862), and finally Corvallis municipal airport (KCVO) weather station. Weather data of the Corvallis Water Bureau (CWB) and Hyslop weather station were obtained from National Oceanic and Atmospheric Administration (NOAA) for 2005 to 2014. Weather data for Corvallis Water Bureau station included only precipitation and minimum and maximum air temperature. The Hyslop weather station and Corvallis municipal airport stations included data for precipitation, minimum and maximum temperature, wind speed, and humidity for the period of 2005 to 2014. SWAT was used to simulate flow and stream temperature throughout the Marys River watershed for the period 2005–2014. This includes the period 2010–2014 when observations for stream temperature were available.

2.3. Stream Temperature Models

2.3.1. Model 1: Linear Regression

The default SWAT stream temperature model uses a linear relation between air temperature and stream temperature developed by Stefan and Preud'homme [12] to calculate stream temperature in the Mississippi River basin, as shown in Equation (1):

$$T_{water} = 5.0 + 0.75 \times T_{air} \tag{1}$$

T_{water} is the average daily water temperature (°C), and T_{water} is the average daily water temperature (°C). Stream temperatures predicted from the above equation will always be higher than air temperature, which is generally a fair assumption for small streams with shallow water depths where stream temperature is primarily controlled by air temperature. However, this may not be necessarily true for streams influenced by snowmelt, surface runoff, and groundwater contributions [10].

2.3.2. Model 2: A Mechanistic Approach Involving Air Temperature and Hydrological Flows

Ficklin et al. [10] developed a mechanistic stream temperature model within SWAT by combining air temperature (heat exchange) and hydrological inputs (flow mixing) including different hydrological parameters, surface runoff, lateral flow, snowmelt, and groundwater contributions. The Ficklin et al. [10] stream temperature model discretizes stream temperature determination into three components: (1) Within the sub-basin, (2) contribution of upstream sub-basins to the targeted sub-basin, and (3) finally heat exchange between air temperature and the stream.

The first part of the stream temperature calculation within the sub-basin (Equation (2)) calculates the local temperature based on a mixing of surface runoff, lateral flow, groundwater, and snowmelt temperatures within the sub-basin flowing to the main stream:

$$T_{w,local} = \frac{\alpha(0.1\,Sub_{snow}) + \beta(T_{gw}\,Sub_{gw}) + \lambda(T_{air,lag}\,Sub_{surq} + Sub_{latq})}{Sub_{wyld}} \tag{2}$$

$T_{air,lag}$ is average daily air temperature with a lag (°C), and α, β, and λ are calibration coefficients that relate the relative contribution of the hydrologic components to local water temperature (dimensionless). Sub_{snow} is the snowmelt contribution in sub-basin (m^3/day), Sub_{gw} is the groundwater contribution in sub-basin (m^3/day), Sub_{surq} is the surface runoff in the sub-basin (m^3/day), Sub_{latq} is the lateral soil flow in sub-basin (m^3/day), and Sub_{wyld} is the water yield in the sub-basin combining all of the above hydrological inputs (m^3/day).

The second part of the Ficklin et al. [10] calculates the effect of upstream sub-basin flow on stream temperature, as shown in Equation (3):

$$T_{waterinitial} = \frac{(T_{w,upstream})(Q_{outlet} - Sub_{wyld}) + (T_{w,local} \times Sub_{wyld})}{Q_{outlet}} \tag{3}$$

$T_{waterinitial}$ is the stream temperature adding the effects of flow within the sub-basin, $T_{w,local}$ was calculated previously, $T_{w,upstream}$ is the water temperature of streams entering the sub-basin (°C), and Q_{outlet} is the stream flow discharge at the outlet of sub-basin (m^3/day).

The final step is to calculate the stream temperature by including the effect of air temperature:

$$T_{water} = T_{waterinitial} + K(T_{air} - T_{waterinitial})(TT) \text{ if } T_{air} > 0$$

$$T_{water} = T_{waterinitial} + K((T_{air} + \varepsilon) - T_{waterinitial})(TT) \text{ if } T_{air} < 0 \tag{4}$$

T_{water} is the final stream temperature of water ($^\circ$C) for a given sub-basin, T_{air} is the average daily air temperature ($^\circ$C), K is the bulk coefficient of heat transfer (1/h), TT is travel time of water through the sub-basin (hour), and finally ε is air temperature addition coefficient (for when air temperature drops below zero).

This mechanistic stream temperature model requires calibration coefficients α, β, γ, k, lag as well as annual groundwater temperatures as inputs. All of the other inputs needed to run the model are provided by SWAT. Groundwater temperature can be estimated from weather data provided as the annual average air temperature, and it is often taken 1–2 $^\circ$C higher than the average annual air temperature [20].

2.3.3. Model 3: A Mechanistic Approach Involving Air Temperature, Hydrological Flows, and Radiative Components

Changes in stream temperature are affected by heat and mass transfers [16] that are dependent on channel morphology, hydrology, and stream vegetation, which provides shading near streams. Vegetation especially helps in cooling temperatures by intercepting and absorbing incoming solar radiation. The mechanistic model introduced in the last section accounts for flow transfer and mixing of various hydrologic components, including sub-basin surface runoff, lateral flow, snowmelt, and ground water, which oftentimes help to reduce temperatures, depending on the season. However, the model utilizes a bulk heat coefficient to account for radiative heat exchange between the air-water interface and does not account for land cover or vegetation near streams. The dependency of the model on the air-water correlation of heat exchange can thereby lead to over-prediction of stream temperatures. Water temperature change related to heat transfer is a function of several sources of radiative heat exchange, as shown in Equation (5):

$$\Phi_{total} = \Phi_{SR} + \Phi_{longwave\text{-}atmosphere} + \Phi_{longwave\text{-}landcover} + \Phi_{convection} + \Phi_{conduction} + \Phi_{evaporation} \tag{5}$$

where, Φ_{total} is the net radiation exchange and is equal to the direct and diffuse solar radiation Φ_{SR} as well as the longwave-atmosphere, longwave-landcover, convection, and evaporation components.

Direct and diffusive solar radiation represent the largest sources of incoming thermal energy into streams. Longwave radiation from different sources also plays a role in increasing and decreasing temperatures: Atmospheric and land-cover longwave radiation add energy to water volumes and increase water temperature, while longwave radiation from the water surface emits radiation from the water surface to the atmosphere to cool streams. Energy loss due to evaporation is considered a larger contributor of decreasing stream temperatures when the required energy is met to change the water phase from liquid to gas. Overall, convection or the air-water interface is considered a very small portion of the total energy budget. Groundwater flux also helps to decrease stream temperatures when added as a thermal input.

These radiative components have been included in the HEATSOURCE model but have not been incorporated explicitly within spatially distributed watershed models such as SWAT. Each of the components are defined as follows.

$$\Phi_{SR} = H_{day} - 0.5 \times H_{day}(1 - e^{-k \times LAI}) \tag{6}$$

Φ_{SR} is the amount of solar radiation reaching the water surface, H_{day} is the incident total solar radiation per day (MJ/m^2·day), k is the light extinction coefficient, and LAI is the leaf area index. Atmospheric solar radiation is calculated as follows:

$$\Phi_{longwave\text{-}atmosphere} = 0.96 \times \varepsilon_{atm} \times \sigma \times (T_{air} + 273.15)^4 \tag{7}$$

$\Phi_{longwave\text{-}atmosphere}$ is the longwave radiation emitted from the atmosphere, ε_{atm} is the emissivity of the atmosphere (unitless), σ is the Stefan-Boltzmann constant (MJ·m^{-2}·day^{-1}·K^{-4}), and T_{air} is the average air temperature per day (°C). The atmospheric solar radiation depends on the emissivity of the atmosphere (ε_{atm}), in which the percent of cloudiness and type of landuse can increase or decrease atmospheric emissivity. The emissivity of the atmosphere is calculated as $\varepsilon_{atm} = 0.767 \times (e_a)^{\frac{1}{7}}$, where ea is the vapor pressure of air (mbar) (i.e., H $\times$ e$_s$), e$_s$ is the saturation vapor pressure (mbar) (i.e., $6.1275 \times e^{\frac{17.27 \times T_{air}}{237.3 + T_{air}}}$), and H is the relative humidity (unitless).

The longwave radiation emitted from landcover $\Phi_{longwave\text{-}landcover}$ is dependent on the view to the sky θ_{VTS} (unitless) and can be calculated as follows:

$$\Phi_{longwave\text{-}landcover} = 0.96 \times (1 - \theta_{VTS}) \times 0.96 \times \sigma \times (T_{air} + 273.15)^4 \tag{8}$$

The last component of longwave solar radiation is the water surface longwave:

$$\Phi_{longwave\text{-}water\ surface} = \varepsilon_w \times \sigma \times (T_s + 273.15)^4 \tag{9}$$

$\Phi_{longwave\text{-}water\ surface}$ surface is the longwave radiation emitted from the water surface, ε_w is the water emissivity taken as 0.97, and T_s is the stream temperature (°C). The evaporation from the water surface is the most effective component in decreasing the thermal energy stored in water:

$$\Phi_{evaporation} = \rho \times L_e \times E \tag{10}$$

ρ is the density of water (kg/m^3), L_e is the latent heat of vaporization (MJ/kg), which is calculated as $L_e = 2.501 - 2.361 \times 10^{-3} \times T_{air}$. E is the evaporation rate of the water surface (m/day) and is calculated using a mass transfer method: $f(w) \times (e_{sw} - e_{aw})$, where $f(w)$ is the wind function $a + b \times w$ that depends on coefficients a and b (mbar^{-1}) and the wind speed w (m/s) measured 2 m above the water surface [16]. Finally, e_{sw} is the saturation vapor pressure of water (mbar), and e_{aw} is the vapor pressure of water (mbar).

The convection radiation component is calculated using the previously calculated evaporative flux $\Phi_{evaporation}$ and Bowen's ratio BR:

$$\Phi_{convection} = BR \times \Phi_{evaporation} \tag{11}$$

Here, BR is unitless (i.e., $0.00061 \times P_A \times \frac{T_{water} - T_{air}}{(e_{sw} - e_{aw})}$), and P_A is the adiabatic air pressure (mbar) [i.e., $1013 - 0.1055 \times z$]. z is the measurement height in meters [i.e., >zd + zo]; zd is the zero-plane displacement (m) [i.e., $0.7 \times H_{Lc}$], zo is the roughness height = $0.1 \times H_{Lc}$, and H_{Lc} is the height of emergent vegetation (m).

The change of stream temperature due to thermal energy flux is calculated as follows:

$$T_{w-TD} = \frac{\Phi_{total}}{\rho_w \times C_w \times d_w} \tag{12}$$

Here, $T_{w-TD}(\frac{°C}{day})$ is the temperature change generated from the thermal components, Φ_{total} is the net driver (MJ/m^2·day), C_w (MJ/kg·C) is the specific heat capacity of water, and d_w (m) is the depth of water in the channel, which is estimated by the SWAT model.

The final stream temperature is calculated by replacing the second term in Equation (4) by the new generated stream temperature, as follows:

$$T_{water} = T_{waterinitial} + T_{w - TD} \tag{13}$$

Note that, as mentioned above, the majority of the components are calculated within the SWAT model automatically.

Overall, this model is useful because it utilizes all of the distributed information (mechanistically simulated via SWAT) regarding stream temperature and its relation to hydrologic components within a networked watershed, yet it also explicitly incorporates radiative energy exchange at the surface of the stream.

2.4. Model Calibration/Validation Methodology

SWAT was used to simulate daily hydrologic discharge at each of the sub-basins within the Marys River watershed from 2010–2014 in order to match observed discharge and stream temperature data. The SWAT model was manually calibrated for stream flow between 2010–2014 using the United States Geological Survey (USGS) (14171000) Philomath flow gauge, which is located 6.7 km southwest of where the Marys River meets the Willamette River, covering a 394 km^2.

The model was only calibrated without validation due to the availability of observations for flow since the available observations only included a period of less than 10 years. Based on the data availability, the model was calibrated for the period of January 2010 to December 2014.

The Nash Sutcliffe efficiency (NSE; Nash and Sutcliffe (1970)) criterion Equation (14) and Pearson's product moment correlation coefficient (1999; Equation (15)) were used to evaluate hydrologic model efficiency:

$$NSE = 1 - \frac{\sum_{i=1}^{n} (O_i - S_i)^2}{\sum_{i=1}^{n} (O_i - O_{avg})^2} \tag{14}$$

$$R^2 = \left(\frac{\sum_{i=1}^{n} (S_i - S_{avg})(O_i - O_{avg})}{\left[\sum_{i=1}^{n} (S_i - S_{avg})^2\right]^{0.5} \left[\sum_{i=1}^{n} (O_i - O_{avg})^2\right]} \right)^2 \tag{15}$$

O is the observed value, S is the model prediction, O_{avg} is the overall observed mean, and S_{avg} is the overall simulated mean. The NSE values range from $-\infty$ to one; a NSE value of less than 0.5 designates an 'unsatisfactory' model, while a NSE value above 0.75 is considered a 'very good' model [21]. R^2 values range from zero to one, with zero indicating a nonlinear relationship between the observed and predicted value and one indicating a perfect fit and a linear relationship between the observed and simulated variables.

The stream temperature models were manually calibrated using root mean square error (RMSE) values as well as percent bias (PBIAS). According to Chai et al. [22], RMSE is widely used as a statistical metric tool to assess performance of models. RMSE can be calculated as follows:

$$RMSE = \sqrt{\frac{\sum_{i=1}^{n} (O_i - S_i)^2}{n}} \tag{16}$$

Where O is observed value, S is model predicted value, and n is the total number of the points. A RMSE value of zero indicates a perfect fit. According to Singh et al. [23], values of RMSE less than half of the standard deviation of the measured data can be taken as acceptable for the model evaluation.

In addition to RMSE, PBIAS was also used to evaluate the models. PBIAS measures the average tendency of the simulated data to be larger or smaller than their observed counterparts [24]. Zero is the optimal value of PBIAS, and a low absolute value implies an accurate model. According to

Gupta et al. [24], positive PBIAS values indicate a model underestimation bias, and negative PBIAS values indicate a model overestimation bias. PBIAS can be calculated as follows:

$$\text{PBIAS} = \frac{\sum_{i=1}^{n} (O_i - S_i) \times 100}{\sum_{i=1}^{n} (O_i)} \tag{17}$$

A set of seven calibration parameters were selected and manually modified to calibrate SWAT for hydrology (Table 1), and five parameters were manually modified to calibrate SWAT for stream temperature (Table 2). Observed daily stream temperature data corresponding to SWAT's sub-basins 8, 15, and 17 were available from 2010 through 2014, while observations for sub-basin 36 extended from 2011 through 2014.

Table 1. Streamflow calibration parameters.

Parameter	Name	File	Range	Calibration Value
CANMX	Maximum canopy storage (mm H_2O)	HRU		+25 for FRSE, FRSE
SMFMX	Melt factor for snow on 21 June (mm H_2O/C-day)	BSN	0–10	8
SMFMN	Melt factor for snow on 12 December (mm H_2O/C-day)	BSN	0–10	1
LAT_TTIME	Lateral flow travel time (days)	HRU		+5
CH_K2	Effective hydraulic conductivity in main channel alluvium (mm/h)	RTE	0–150	+6
GWQMN	Threshold depth of water in the shallow aquifer required for return flow to occur (mm H_2O)	GW	0–5000	* 2.5
CN2	Initial SCS runoff curve number for moisture condition II	MGT	0–100	* 0.978
ESCO	Soil evaporation compensation factor	BSN or HRU	0–1	−0.25

* Values are percentages of the original values.

Table 2. Basin-wide stream temperature calibration parameters.

Parameter	Name	Range	Calibrated Values
α	Coefficient influencing snowmelt temperature contributions (unitless)	0–1	1.0
β	Coefficient influencing groundwater temperature contributions (unitless)	0–1	0.97
λ	Coefficient influencing surface and lateral flow temperature contributions (unitless)	0–1	1.0
K	Bulk coefficient of heat transfer (1/h)	0–1	0.025
Lag	Average air temperature lag (days)	0–14	6

To compare the differences in model performance, kernel density estimates were calculated using R software. This nonparametric technique is similar to using histograms to highlight the differences between model simulations and observed data for the three tested models.

3. Results and Discussion

3.1. Hydrology Calibration in SWAT

As mentioned previously, SWAT was used to simulate daily hydrologic discharge at each of the sub-basins within the Marys River watershed for 2005–2014, which included 2010–2014, when observations for stream temperature were also available. The NSE and R^2 values of the default model's simulated flow (no calibration) were −0.37 and 0.50, respectively, indicating an unsatisfactory model. From Figure 2, it is clear that the uncalibrated model overpredicts the peaks during storm events. Also, the model's responses to each rain event are very rapid, and the water loss rates are excessive, which results in zero flow for late summer periods.

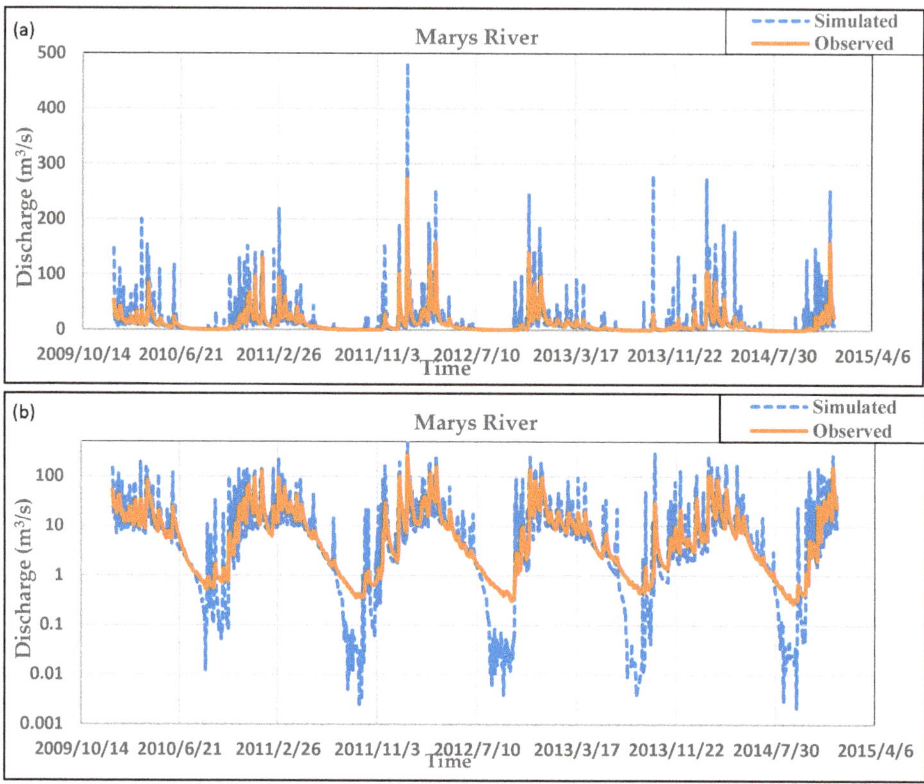

Figure 2. (**a**) Uncalibrated daily SWAT discharge simulations compared with observations (2010–2014) from the USGS (14171000) Philomath flow gage. (**b**) Uses a log scale to emphasize low-flow conditions.

Manual calibration resulted in model improvement of NSE values from −0.37 to 0.72. This designates a 'good' model according to Moriasi et al. [21], since it is >0.65. The R^2 value of the model increased to 0.80. Figure 3 shows that the calibrated model matches both the base flow and peaks well, whereas all of the peaks were mainly over predicted by the default model. The default model could not capture the very low stream flows for some summer days and resulted in zero flow (Figure 2), but the calibrated model (Figure 3) fixed the low-flow problem and improved the results. Figure 2 shows that the maximum simulated peak for the uncalibrated model was around 500 m^3/s, whereas the calibrated model (Figure 3) reduced this value to match the peak observed around 300 m^3/s. The rapid response of the main channel to any storm event led to the over-prediction of peaks even for small rain storms in the uncalibrated model. Also, the lateral flow travel time parameter helped to slow the response to storm events and smoothed the hydrograph. The other major problem was associated with the excessive loss of water in a short period of time after a rapid response to any storm; water in the uncalibrated model was lost instantaneously and therefore resulted in zero flow for late summer days. The effective hydraulic conductivity in the main channel (CH_K2) was used to prevent the excessive loss from the main channel and managed to eliminate the zero flow days. Also, this parameter helped to smooth and eliminate the transient fluctuations in the hydrograph. The SCS curve number for moisture conditions II (CN2) was used to reduce or increase the simulated peaks to match the observed hydrograph. The canopy interception parameter for specified land cover types (CANMX) as well as the minimum and maximum snowmelt factors (SMFMN, SMFMX) all helped to increase the evaporation rate. The high surface runoff surge was decreased using the soil evaporation compensation factor

(ESCO), the SCS curve number for moisture conditions II (CN2), and the threshold depth of water in the shallow aquifer (GWQMN).

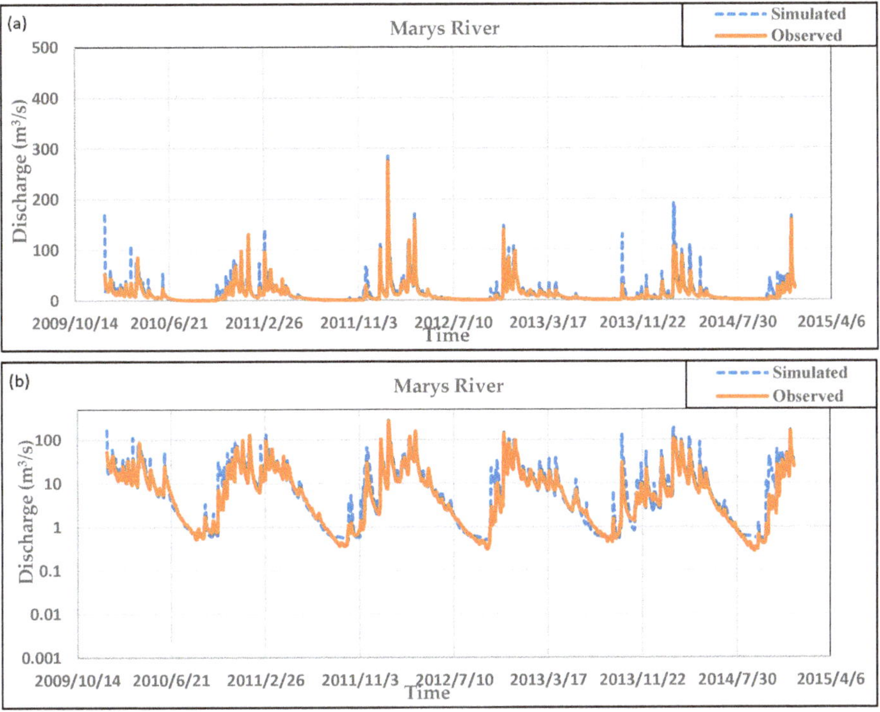

Figure 3. (**a**) Manually calibrated daily Soil and Watershed Assessment Tool (SWAT) discharge simulations compared with observations (2010–2014) from the United State Geological Survey (USGS) (14171000) Philomath flow gage. (**b**) Uses a log scale to emphasize low-flow conditions.

3.2. Stream Temperature Calibration in SWAT

After a satisfactory hydrologic calibration was performed, manual calibration was performed for two of the three stream temperature models that will be tested in the Marys River using SWAT. The first model is the linear regression model from Stefan and Preud'homme [12] and was not calibrated. The second model is the Ficklin et al. [10] model, and the third model is our extension to the Ficklin et al. [10] model. As shown in Table 2, we manually calibrated five parameters to best match the second and third models to the stream temperature observations between 2010 and 2014. Note that only summer stream temperature observations were available and that only a single set of calibration parameters were used for the entire watershed.

3.3. Stream Temperature Model Comparison

After satisfactory hydrologic and stream temperature calibrations were performed, three models were tested to simulate stream temperature in the Marys River using SWAT. The first was the linear regression model from Stefan and Preud'homme [12], the second the Ficklin et al. [10] model, and the third is our extension to the Ficklin et al. [10] model that incorporates HEATSOURCE radiative forcing components to the Ficklin et al. [10] model. For the remainder of this paper, these models will be referred to as Model 1, Model 2, and Model 3, respectively.

Table 3 shows RMSE and PBIAS results for these three models using daily data. Kernel density estimates for the differences between simulated and observed stream temperatures for the four sub-basins are plotted in Figure 4.

Table 3. Comparison of root mean square error (RMSE) and percent bias (PBIAS) values for the three tested stream temperature models (daily).

Sub-Basin	Period	RMSE			PBIAS (%)		
		Model 1	Model 2	Model 3	Model 1	Model 2	Model 3
8	2010–2014	3.74	2.18	2.36	23.2	6.9	2.3
15	2010–2014	3.46	1.96	1.96	21.2	3.9	−0.5
17	2010–2014	2.6	1.88	2.72	13.4	−2.6	−8.3
36	2011–2014	2.85	2.28	3.12	13.6	−2.2	−1.0

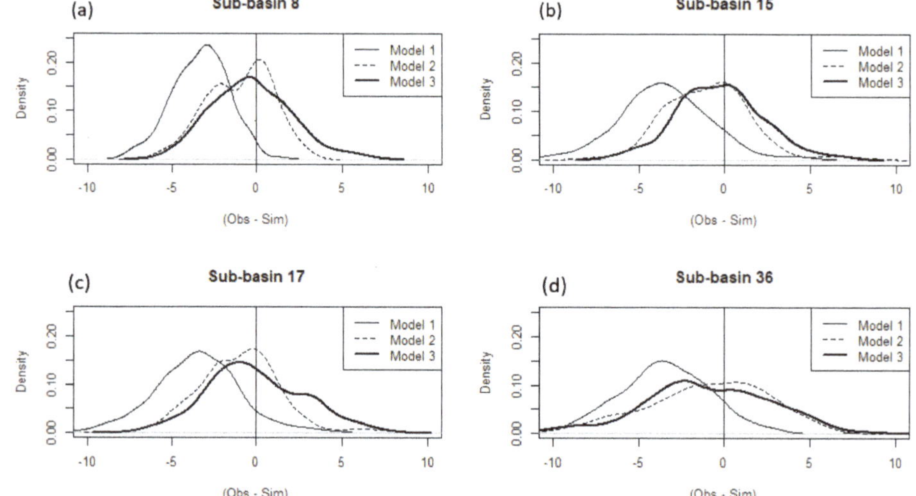

Figure 4. Kernel density estimates for the differences between observed and simulated stream temperatures as calculated using the three models for four sub-basins (**a–d**). The vertical line at zero indicates simulations that exactly match the observed data.

In general, Model 1 performed the worst among all three models (see Table 3 and Figure 4), and Model 2 outperformed Models 1 and 3 for all sub-basins. Model 1 consistently overestimated stream temperatures for all of the sub-basins. This is apparent in Figure 4, where the distributions of the differences between the simulated and observed temperatures are shifted from zero for Model 1. The coefficients of Model 1 guarantee that the simulated stream temperature will be above average daily air temperature values when the average air temperature is less than 20 °C. Therefore, inclusion of cold water from groundwater or upstream sources is not captured in this model, thus resulting in overestimations. Model 2, which is the calibrated Ficklin et al. [10] model, showed improvements compared to the linear model (Model 1) in both Table 3 and Figure 4, which agrees with previous studies (Barnhart et al., 2014; Ficklin et al., 2012; Ficklin et al., 2014) [10,13,14]. This is presumably because Model 2 incorporates hydrologic components, including groundwater upstream temperatures, in addition to an air-heat exchange transfer coefficient. Model 3, which replaced the simple air-heat exchange transfer coefficient from Model 2 with explicitly calculated radiative components, shows similar distributions with Model 2 (Figure 4), yet the performance values of Model 2 are generally better. This is likely due to the high variability of the radiative components included in Model 3, which will be discussed further in the next section.

3.4. Land Cover Effects on Stream Temperature

We now compare Models 2 and 3 to demonstrate that the incorporation of radiative components is able to simulate the influence of land use and land cover on stream temperatures (e.g., forested vs. agricultural regions). Stream temperature simulations for two sub-basins—a forested area (sub-basin 8) and an area dominated by agriculture with low vegetation (sub-basin 36)—are shown in Figure 5.

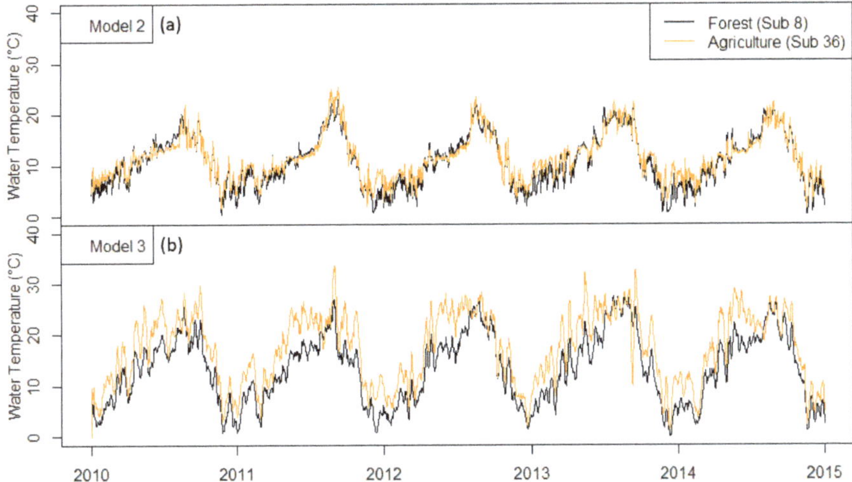

Figure 5. Stream temperatures simulated using the Ficklin et al. [10] model (Model 2; panel (**a**)) and the extended version that includes radiative components (Model 3; panel (**b**)). Model 2 does not capture changes in stream temperature associated with land cover type, while Model 3 simulates increased temperatures for agricultural regions.

Model 2 simulates nearly identical stream temperatures for both forested and agricultural sub-basins. Conversely, by including the various radiative components into the model, Model 3 simulates consistently increased stream temperatures associated with the agricultural sub-basin. This is especially apparent during the early summer months. To examine differences in the models further, Figure 6 compares the net radiative components of Model 3 Equation (5) with the K-component second term in Equation (4) of Model 2.

Figure 6 shows that the net radiation as calculated using Model 3 (orange lines) changes according to the primary land use cover for a given sub-basin in SWAT. For example, agricultural sub-basins have larger incoming (positive) radiation contributions that help to increase stream temperatures, whereas forested areas have small or negative radiative effects, depending on the season, due to increased LAI, reduced solar radiation reaching the water surface, and evaporative fluxes. Model 2 uses a bulk coefficient of heat transfer in the second term of Equation (4). This reflects a convection component of the net radiative balance, but it does account for the other radiative energy terms, including solar radiation, atmospheric longwave, land surface longwave, water surface longwave, and evaporation. Therefore, it is not able to capture cover-related differences in net radiation and therefore changes in stream temperature due to landscape and land use changes.

Figure 6 also shows that the net radiative driver as calculated in Model 3 has much higher variability than the K-component used in Model 2. We found that the high variability (i.e., noise) is mainly due to SWAT's estimation of solar radiation as well as the evaporation calculations, which are not explicitly accounted for in Model 2. This likely led to the reduced model performance exhibited when comparing model simulations to observed data.

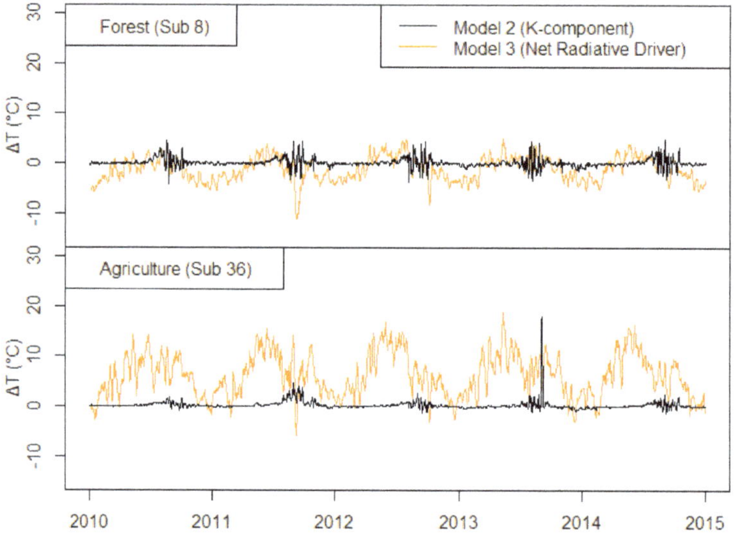

Figure 6. Comparison of the thermal energy impacts on stream temperatures for Models 2 and 3 in both primarily forested and agricultural sub-basins.

Figure 7 compares these data further by plotting the observed stream temperature as well as simulations using Models 2 and 3 for both the forested (sub-basin 8) and agricultural (sub-basin 36) sub-basins.

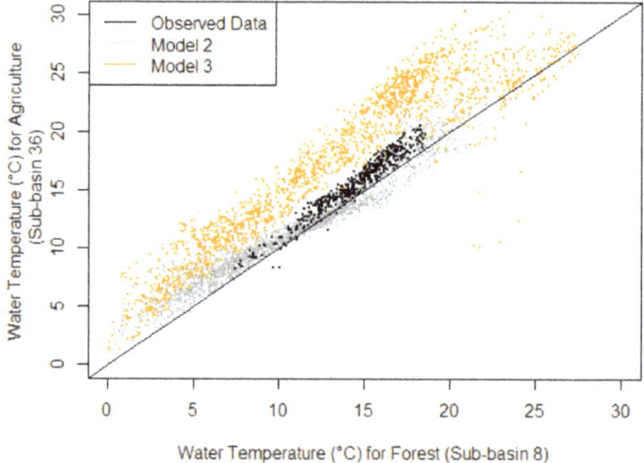

Figure 7. Comparison of observed and simulated stream temperatures for agricultural and forested sub-basins. The observed data (black) show a consistent increase in stream temperatures for the agricultural sub-basin (sub-basin 36). Model 2 does not capture consistent changes in stream temperature associated with land cover type. Model 3 simulates increased temperatures for the agricultural sub-basin, but it overestimates the effect compared to the observed data.

The mean (standard deviation) differences between the stream temperature simulations for the forested sub-basin (8) and the agricultural sub-basin (36) were $-0.68\,^{\circ}$C ($1.30\,^{\circ}$C) and $-4.1\,^{\circ}$C ($2.34\,^{\circ}$C) for Models 2 and 3, respectively. These simulations can be compared with the difference between the observed data for the forested and agricultural sub-basins (8 and 36): $-1.07\,^{\circ}$C ($0.63\,^{\circ}$C). Note that Model 2 gives estimates that are closer to the observed values than Model 3; this is not surprising since the RMSE values for Model 2 were lower than the values for Model 3 (Table 3). However, Model 2 is symmetric about the 1:1 line shown in Figure 7 and does not capture the positive bias of agricultural stream temperatures that is shown by the observed values as well as in the simulated values of Model 3. Model 3 captures the increases in stream temperature associated with the lack of forest cover in the agricultural sub-basin, yet Model 3 overpredicts this effect, and the variation in the simulated values are much greater than Model 2 or the observed data.

Overall, Model 3 may be useful for simulating the watershed-scale impacts of land use conversion from forest to agriculture on stream temperatures; however, model improvement—potentially through improved calibration—is needed to better match observed data. In addition, while Model 3 is able to simulate impacts of land use on stream temperature, this advantage is produced with the trade-off that radiative components exhibit much higher variability, and this noisy fluctuation (which can be seen directly from Figures 5–7) further decreases RMSE values (Table 3).

4. Conclusions

This study sought to explicitly incorporate radiative forcing components into an existing mechanistic, semi-distributed stream temperature model Ficklin et al. [10] using SWAT. Ultimately, stream temperature is controlled by different climate components including humidity, wind speed, evaporation, and solar radiation besides air temperature and is also heavily dependent on hydrologic components including surface flow, groundwater, and snow melt processes. Our new model leverages the Ficklin et al. [10] stream temperature model, which accounts for hydrological and climatological components and discretizes the prediction of stream temperature into three parts: (1) Streamflow within a sub-basin, (2) contributing upstream sub-basin hydrologic components, and (3) accounting for the heat exchange between air and water surface, which can be described as a convection term. The extended model replaces the convection K-component term within the Ficklin et al. [10] model with a more comprehensive characterization of radiative energy terms, including solar radiation, atmospheric longwave, land surface longwave, water surface longwave, evaporation, and finally convection drivers. The extended model was used along with the Ficklin et al. [10] model and the linear regression model to simulate stream temperatures within agricultural, forested, and mixed sub-basins within the Marys River watershed. Results showed that all models performed reasonably well, and the Ficklin et al. [10] model outperformed the others. However, the extended model was capable of simulating differences between stream temperatures associated with agricultural and forested watersheds that reflected observed data, although the differences were overestimated. The reduced performance of the extended model that included radiative components might be able to be improved by further calibration; yet, the high variability of the radiative terms is also limiting. For example, the model relies on incoming solar radiation as well as wind velocity, which are difficult to represent over large spatial scales and feature high variability. Alternative formulations of the radiative components should be considered in future work. Bogan et al. [1] suggests using a shading factor instead of a leaf area index, which may lead to more accurate results, assuming shading information is available for the watershed of interest. In addition, alternative estimations for solar radiation, or perhaps any of the radiative energy terms, could improve the model and should be pursued. Overall, incorporating radiative components into the Ficklin et al. [10] stream temperature provides a new mechanism for simulating the effects of alternative land uses on stream temperature within SWAT. This will be especially useful for land managers and decision makers when considering alternative land management scenarios and conservation strategies using SWAT.

Water **2018**, *10*, 1143

Author Contributions: Conceptualization, M.B.-S.; Methodology, M.B.-S.; Software, M.M. and B.B.; Validation, M.M. and B.B.; Formal Analysis, M.M.; Writing—Original Draft Preparation, M.M. and B.B.; Writing—Review & Editing, M.B.-S. and D.F.; Visualization, M.M. and B.B.; Supervision, M.B.-S.

Funding: This research received no external funding.

Acknowledgments: The authors thank reviewers for their comments that helped in preparing a better version for publication. Also, we thank the editorial team of Water for proofreading and preparing the paper for publication. The views expressed in this article are those of the authors and do not necessarily represent the views or policies of the U.S. Environmental Protection Agency. Any mention of trade names, products, or services does not imply an endorsement by the U.S. Government or the U.S. Environmental Protection Agency. The EPA does not endorse any commercial products, services, or enterprises.

Conflicts of Interest: We declare no conflicts of interest.

References

1. Bogan, T.; Mohseni, O.; Stefan, H.G. Stream temperature-equilibrium temperature relationship. *Water Resour. Res.* **2003**, *39*, 149–163. [CrossRef]
2. Knouft, J.H.; Darren; Ficklin, L. The Potential Impacts of Climate Change on Biodiversity in Flowing Freshwater Systems. *Annu. Rev. Ecol. Evol. Syst.* **2017**, *48*, 111–133. [CrossRef]
3. Van Vliet, M.T.; Franssen, W.H.; Yearsley, J.R.; Ludwig, F.; Haddeland, I.; Lettenmaier, D.P.; Kabat, P. Global river discharge and water temperature under climate change. *Global Environ. Chang.* **2013**, *23*, 450–464. [CrossRef]
4. Eaton, J.; Scheller, R. Effects of climate warming on fish thermal habitat in streams of the United States. *Limnol. Oceanogr.* **1996**, *41*, 1109–1115. [CrossRef]
5. Kaushal, S.S.; Likens, G.E.; Jaworski, N.A.; Pace, M.L.; Sides, A.M.; Seekell, D.; Wingate, R.L. Rising stream and river temperatures in the United States. *Fron. Ecol. Environ.* **2010**, *8*, 461–466. [CrossRef]
6. Sherwood, Courtney. Thousands of Salmon Die in Hotter-than-Usual Northwest Rivers. *U.S. Reuters*, 28 July 2015.
7. Mayer, P.M.; Todd, A.H.; Okay, J.A.; Dwire, K.A. Introduction to the featured collection on riparian ecosystems & buffers. *JAWRA J. Am. Water Resour. Assoc.* **2010**, *46*, 207–210.
8. Brown, G.W. The Alsea watershed study. In *Pacific Logging Congress Loggers Handbook*; Pacific Logging Congress Loggers: Portland, OR, USA, 1972.
9. Bond, R.M.; Stubblefield, A.P.; van Kirk, R.W. Sensitivity of Summer Stream Temperatures to Climate Variability and Riparian Reforestation Strategies. *J. Hydrol. Reg. Stud.* **2015**, *4*, 267–279. [CrossRef]
10. Ficklin, D.L.; Luo, Y.; Stewart, I.T.; Maurer, E.P. Development and application of a hydroclimatological stream temperature model within the Soil and Water Assessment Tool. *Water Resour. Res.* **2012**, *48*, W01511. [CrossRef]
11. Arnold, J.G.; Moriasi, D.N.; Gassman, P.W.; Abbaspour, K.C.; White, M.J.; Srinivasan, R.; Santhi, C.; Harmel, R.D.; van Griensven, A.; Van Liew, M.W.; et al. SWAT: Model use, calibration, and validation. *Trans. ASABE* **2012**, *55*, 1491–1508. [CrossRef]
12. Stefan, G.H.; Preud'Homme, B.E. Stream Temperature Estimation from Air Temperature1. *JAWAR J. Am. Water Resour. Assoc.* **1993**, *29*, 27–45. [CrossRef]
13. Barnhart, B.L.; Whittaker, G.W.; Ficklin, D.L. Improved Stream Temperature Simulations in SWAT Using NSGA-II for Automatic Multi-Site Calibration. *Trans. ASABE* **2014**, *57*, 517–530.
14. Ficklin, D.L.; Barnhart, B.L.; Knouft, J.H.; Stewart, I.T.; Maurer, E.P.; Letsinger, S.L.; Whittaker, G.W. Climate change and stream temperature projections in the Columbia River basin: Habitat implications of spatial variation in hydrologic drivers. *Hydrol. Earth Syst. Sci.* **2014**, *18*, 4897–4912. [CrossRef]
15. Du, X.; Shrestha, N.K.; Ficklin, D.L.; Wang, J. Incorporation of the equilibrium temperature approach in a Soil and Water Assessment Tool hydroclimatological stream temperature model. *Hydrol. Earth Syst. Sci.* **2018**, *22*, 2343–2357. [CrossRef]
16. Boyd, M.; Kasper, B. *Analytical Methods for Dynamic Open Channel Heat and Mass Transfer: Methodology for Heat Source Model*; Version 7.0.; Watershed Sciences Inc.: Portland, OR, USA, 2003.
17. Lee, K.K.; Risley, J.C. *Estimates of Ground-Water Recharge, Base Flow, and Stream Reach Gains and Losses in the Willamette River Basin, Oregon*; US Department of the Interior, US Geological Survey: Portland, OR, USA, 2002.
18. Neitsch, S.L.; Arnold, J.G.; Kiniry, J.R.; Williams, J.R. *SWAT User Manual (Version 2009), Texas Water Resources Institute Technical Report*; Texas A&M University System: College Station, TX, USA, 2005.

19. USDA: Geospatial Data Gateway. Available online: https://datagateway.nrcs.usda.gov (accessed on 15 February 2015).
20. Todd, D.K. *Groundwater. Hydrology*, 2nd ed.; John Willey and Sons: New York, NY, USA, 1980.
21. Moriasi, D.N.; Arnold, J.G.M.; Van Liew, W.; Bingner, R.L.; Harmel, R.D.; Veith, T.L. Model Evaluation Guidelines for Systematic Quantification of Accuracy in Watershed Simulations. *Trans. ASABE* **2007**, *50*, 885–900. [CrossRef]
22. Chai, T.; Draxler, R.R. Root Mean Square Error (RMSE) or Mean Absolute Error (MAE)?—Arguments against Avoiding RMSE in the Literature. *Geosci. Model Dev.* **2014**, *7*, 1247–1250. [CrossRef]
23. Singh, J.; Knapp, H.V. Hydrologic modeling of the Iroquois River watershed using HSPF and SWAT. *J. Am. Water Resour. Assoc.* **2005**, *41*, 343–360. [CrossRef]
24. Gupta, H.; Sorooshian, S.; Yapo, P. Status of Automatic Calibration for Hydrologic Models: Comparison with Multilevel Expert Calibration. *J. Hydrol. Eng.* **1999**, *4*, 135–143. [CrossRef]

Article

Improved Soil Temperature Modeling Using Spatially Explicit Solar Energy Drivers

Jonathan J. Halama [1,2,*], Bradley L. Barnhart [1], Robert E. Kennedy [2], Robert B. McKane [1], James J. Graham [3], Paul P. Pettus [1], Allen F. Brookes [1], Kevin S. Djang [4] and Ronald S. Waschmann [1]

[1] Western Ecology Division, National Health and Environmental Effects Research Laboratory, U.S. Environmental Protection Agency, Corvallis, OR 97330, USA; Barnhart.Brad@epa.gov (B.L.B.); Mckane.Bob@epa.gov (R.B.M.); Pettus.Paul@epa.gov (P.P.P.); Brookes.Allen@epa.gov (A.F.B.); Waschmann.Ron@epa.gov (R.S.W.)

[2] College of Earth and Atmospheric Sciences, Oregon State University, Corvallis, OR 97331, USA; rkennedy@coas.oregonstate.edu

[3] Environmental Science & Management, Humboldt State University, Arcata, CA 95521, USA; james.graham@humboldt.edu

[4] Inoventures (LLC), Western Ecology Division, National Health and Environmental Effects Research Laboratory, c/o U.S. Environmental Protection Agency, Corvallis, OR 97330, USA; Djang.Kevin@epa.gov

[*] Correspondence: Halama.Jonathan@epa.gov or halamaj@oregonstate.edu; Tel.: +1-541-737-6332

Received: 31 August 2018; Accepted: 4 October 2018; Published: 9 October 2018

Abstract: Modeling the spatial and temporal dynamics of soil temperature is deterministically complex due to the wide variability of several influential environmental variables, including soil column composition, soil moisture, air temperature, and solar energy. Landscape incident solar radiation is a significant environmental driver that affects both air temperature and ground-level soil energy loading; therefore, inclusion of solar energy is important for generating accurate representations of soil temperature. We used the U.S. Environmental Protection Agency's Oregon Crest-to-Coast (O'CCMoN) Environmental Monitoring Transect dataset to develop and test the inclusion of ground-level solar energy driver data within an existing soil temperature model currently utilized within an ecohydrology model called Visualizing Ecosystem Land Management Assessments (VELMA). The O'CCMoN site data elucidate how localized ground-level solar energy between open and forested landscapes greatly influence the resulting soil temperature. We demonstrate how the inclusion of local ground-level solar energy significantly improves the ability to deterministically model soil temperature at two depths. These results suggest that landscape and watershed-scale models should incorporate spatially distributed solar energy to improve spatial and temporal simulations of soil temperature.

Keywords: soil temperature; solar energy; watershed model; landscape scale; VELMA

---

1. Introduction

Soil temperature affects several key ecosystem properties. Through surface runoff and subsurface groundwater transport, soil temperatures can lead to increased stream temperatures, which in turn impact salmonid and other fish habitats [1]. Soil temperatures mediate rates of biogeochemical transformations in soils, strongly influencing local to global-scale patterns in the cycling, retention and loss of carbon and nutrients from ecosystems [2]. Seasonal soil temperature trends can shift photosynthetic recovery timing and therefore impact overall net primary production (NPP) [3]. Such soil temperature effects are subject to modification by physical landscape factors, such as object shading, slope aspect and thermal isolation from a detritus layer or snow pack [4].

Mechanistic watershed models such as Visualizing Ecosystem Land Management Assessments (VELMA) (2.0, U.S. Environmental Protection Agency—Western Ecology Division, Corvallis, OR, USA) [5], Soil and Water Assessment Tool (SWAT) (2009, Texas A&M, College Station, TX, USA) [6], Regional Hydro-Ecologic Simulation System (RHESSys) (2004, University of California, Santa Barbara, CA, USA) [7], and Hydrologic Simulation Program—Fortran (HSPF) (11, U.S. Environmental Protection Agency—National Exposure Research Laboratory, Athens, GA, USA) [8] use a mechanistic (as opposed to statistical) approach to model hydrodynamics throughout a watershed using sub-daily or daily time steps. Models such as these utilize equations to simulate hydrologic dynamics and soil moisture by tracking the rate of water transfer based on soil porosity, soil depth and the available precipitation.

Watershed models simulating soil temperature at multiple depths rely on several observed and simulated environmental variables including air temperature, precipitation, soil moisture, soil depth, and physical soil properties. For national and regional-scale modeling purposes, climate data can often be obtained from various governmental agencies such as the National Oceanic and Atmospheric Administration (NOAA), Natural Resources Conservation Service (NRCS) and others that maintain large-scale networks of climate monitoring stations such SNOTEL and SCAN [9]. At more local scales, climate data collection tends to focus on site-specific requirements, such as municipal airports, Long Term Ecological Research Stations and university research forests [10–13]. This can result in data limitations for spatially explicit models [14]. Several groups process the site data to produce spatial datasets at various spatial scales (e.g., Parameter-elevation Regressions on Independent Slopes Model (PRISM) and Daily Surface Weather Data (Daymet)) [15,16]. Soil column properties can be acquired through field work or obtained by utilizing soil datasets (e.g., State Soil Geographic (STATSGO) and Soil Survey Geographic (SURGO)) [17].

While climate and soil properties influence soil temperatures, solar energy is the most significant environmental variable influencing soil temperature. There are existing spatial models that account for solar energy inputs at a local or stream reach scale: SHADE2 (1.0, University of Georgia, Athens, GA, USA) [18], HeatSource (8.0, Oregon Department of Environmental Quality, Portland, OR, USA) [19], and iLand (1.0, Seidl and Rammer, Vienna, Austria) [20], utilize small-area representations of solar energy. However, these model's spatial heterogeneity of shade is utilized for quantifying shade along stream reaches or within forest plots; not ground-level irradiance within models representing complete watersheds or landscapes.

Previous methods of incorporating solar energy within model representations of complete watersheds employ one or more proxy variables (e.g., canopy coverage or air temperature), or they simply utilize an average daily global irradiance value. Current environmental mechanistic models (e.g., VELMA, SWAT, RHESSys, HSPF) [5–8] use a global solar irradiance subroutine to calculate the total solar energy (W/m^2) for the entire watershed or for sub-catchments. Due to their simpler ground-level solar energy representations, although these models capture the seasonal pattern of irradiance, they lack a spatially heterogenous representation of topographic and landscape object shading that affects ground-level solar energy levels. Solar energy estimate methods themselves include uncertainty due to environmental variables like cloud fraction and albedo [21]. Additional variables (i.e., aerosol optical properties, cloud asymmetry, water vapor distribution) may have seasonal and regional influence on the accuracy of solar energy estimates [22].

There remains a gap in soil temperature modeling where current approaches utilize global solar energy models that are not capturing local energy interactions. Finer-scale spatially distributed estimates of ground-level shade or solar energy could provide improved soil temperature estimates. This paper addresses this gap by incorporating local solar energy data within the soil temperature subroutine of an ecohydrological watershed model to determine its effect on simulated soil temperature predictions at multiple depths.

To demonstrate the utility of linking spatially explicit, ground-level solar energy data with an environmental model, we focus on improving VELMA's soil temperature subroutine by incorporating spatially heterogeneous solar energy as an input driver. First, we discuss a commonly used global solar energy model and soil temperature model that are found in many deterministic watershed models, including VELMA. Next, we present VELMA's original soil temperature subroutine that incorporates spatially explicit inputs of soil moisture and air temperature but does not employ any form of solar energy presentation. Following the original model, we then present VELMA's modified soil temperature subroutine that incorporates spatially explicit inputs of ground-level solar energy, along with the inclusion of soil moisture and air temperature. We compare the predictive skill of the original and new forms of VELMA's soil temperature subroutines using multiple United States Environmental Protection Agency (EPA) Oregon Crest-to-Coast Environmental Monitoring Transect (O'CCMoN) sites. This transect consists of several paired sites of forested and open landscapes [23]. We present results that demonstrate the benefit of including spatially explicit representations of solar energy within watershed-scale models that simulate soil temperature.

2. Materials and Methods

Watershed models typically include solar energy directly or through a proxy variable to facilitate energy requirements needed within subroutine routines. Plant growth models may require a daily input of solar energy reaching the canopy to drive photosynthesis [24]. Stream temperature models predict shifts in water temperature through variables representing landscape shading, water temperatures, and air temperatures, all of which are solar energy proxies. Snowmelt models may need a daily input of solar energy or air temperature to drive snow melt [25]. While all these subroutines rely on solar energy at the earth's surface, mechanistic models generally lack the ability to capture the spatiotemporal dynamics of solar energy reaching the ground post shadowing.

The soil temperature subroutine from VELMA [5] was chosen for testing. VELMA is a spatially distributed watershed model that simulates hydrologic and biogeochemistry processes within a gridded framework under mechanistic cell interactions. Using a gridded framework, VELMA describes each grid cell as having a ground-level surface and four sub-surface voxels representing the landscapes soil strata. Each subsurface voxel is characterized by soil porosity and soil depth. Water transfers at a daily time step through VELMA's voxel framework. Based on water transmission, nutrients and thermal energy migrate through the simulated soil substrate under mechanistic rules.

2.1. Previous Solar and Generalized Soil Modeling Methods

Watershed models often employ a clear-sky solar energy model when direct solar energy units are required. A common approach, and the method used by VELMA, is to calculate the clear-sky solar energy that reaches the earth's surface as in Equation (1), where R is solar irradiance (W/m^2), ecc is the eccentricity correction factor, w is the Earth's constant angular velocity, T is the time frequency, dec is the solar declination, and γ is latitude [26]:

$$R = (24/\pi) \times 4.921 \times ecc \times [wT \times \sin(dec) \times \sin(\gamma) + \cos(dec) \times \sin(wT) \times \cos(\gamma)] \tag{1}$$

The VELMA model uses Equation (1) to describe the amount of solar energy reaching the troposphere under clear sky conditions. This approach does not account for the topographic or object shading that locally reduces ground-level solar energy.

Soil temperature modeling within many watershed models, whether utilizing a gridded representation of the landscape or aggregating to sub-catchment scales, typically uses some version of the Carslaw and Jaeger equation to quantify seasonal variation in soil temperature [27]:

$$T_{soil}(z, d_n) = T_{AA} + A_{surf} \times e^{-z/dd} \times \sin(\omega \times d_n - z/dd) \tag{2}$$

where $T_{soil}(z, d_n)$ is the soil temperature (°C) at depth z (mm) for day of the year d_n, T_{AA} is the average annual soil temperature, A_{surf} is the amplitude of the surface fluctuations, dd is the damping depth (mm), and ω is the angular frequency of the damping oscillations by day (d_n). At $z = 0$, the soil temperature reduces to the following:

$$T_{soil}(0, d_n) = T_{AA} + A_{surf} \times \sin(\omega \times d_n) \tag{3}$$

which is the average soil temperature perturbed by surface temperature fluctuations and reflects seasonal solar patterns. Conversely, at infinite depth, the soil temperature becomes equal to the annual average soil temperature. This formulation provides a relatively simple method for calculating soil temperatures at multiple depths throughout a watershed. However, the model requires specification of soil heat capacity as well as thermal conductivity to correctly specify the amplitude coefficient and the damping depth.

2.2. Soil Temperature Variations Due to Landscape Coverage

Temperature profiles of soils can dramatically vary between a forested versus open environment, even if the sites are located proximally near one another. Two sites can be exposed to very similar climate conditions, though due to forest canopy shading, the forested site will have reduced air temperature and a reduction in solar energy loading upon the soil surface. Figure 1 shows observed 2005 daily soil temperature differences between open (clear-cut harvested) site data minus forested site data for the O'CCMoN Soapgrass field site in Oregon (see Section 2.5.1 for field site locations).

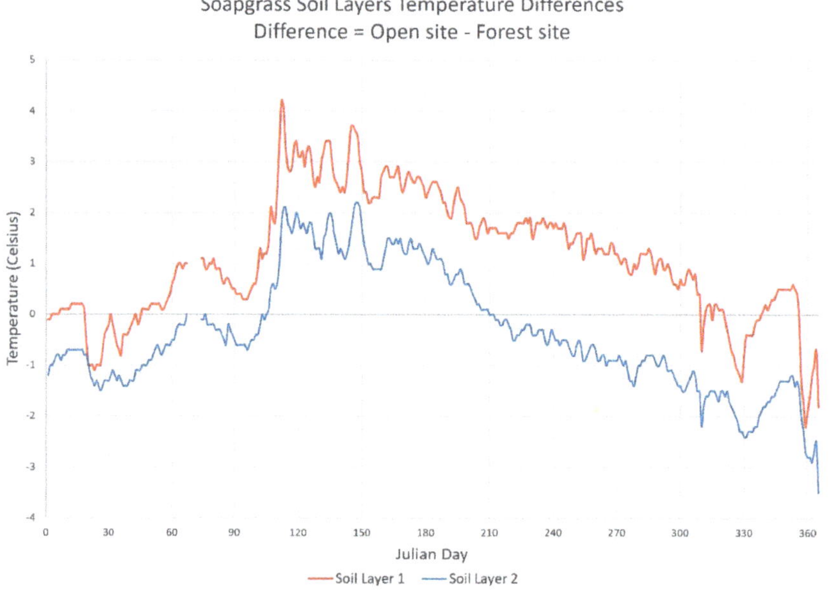

Figure 1. Daily differences in soil temperature (°C) between open (clear-cut) and forest sites at the O'CCMon Soapgrass site. Temperature differences were calculated as the Open Site temperature minus the forest site temperature at each Julian day during the year 2005. Thus, positive values denote days where the open site soil temperature was warmer than the forest site soil temperature. Open minus forest soil temperature differences were calculated for each of the two soil depths, 15 cm (red line) and 30 cm (blue line) below the soil surface. The data gap between Julian days 67 and 74 was due to sensor errors.

Figure 1 highlights the soil temperature differences between open and forested sites. A positive temperature means the open site temperature was warmer than the forest site; conversely, negative temperature means the open site temperature was colder than the forest site. Two main observations should be made here: (1) layer 1 is always warmer than layer 2, and (2) for both soil layers the open site is always warmer in summer than the forest site, but is comparatively colder in winter and especially so in layer 2. The open site is significantly warmer from Julian day 45 through 310 (14 February through 6 November), with a peak difference of 4.2 °C on Julian day 111 (21 April).

Seasonal differences in the warming and cooling of soils in the open and forest sites (Figure 1) certainly reflect changes in air temperature along with some complicating effects associated with inter-site variations in snow pack and associated insulative properties (Figure 2). Other factors undoubtedly also come into play, such as the effects of seasonal changes in soil moisture (dry summers, wet winters) on soil thermal transmissivity.

Nevertheless, observed increases in summer air temperatures for the open site tended to be 2.12 °C warmer than the forest site (Figure 2). The lowest thermal difference was −7.65 °C while the highest thermal difference was 8.0 °C. During the same period, soil layer 1 open site averaged 1.71 °C warmer than the forest site with a minimum of 1.0 °C and a maximum of 2.6 °C. However, soil layer 2 open site averaged −0.16 °C cooler than the forest site with a minimum of −1.0 °C and maximum of 1.3 °C. That is, while air temperature differences are driven by differences in solar radiation, in the open site there is clearly an additional direct effect of solar radiation on heat transfer to the ground surface and consequent warming of the soil column. This observation underscores the importance of quantifying the direct effect of solar radiation on soil temperature, in combination with effects of soil moisture and other factors mentioned above.

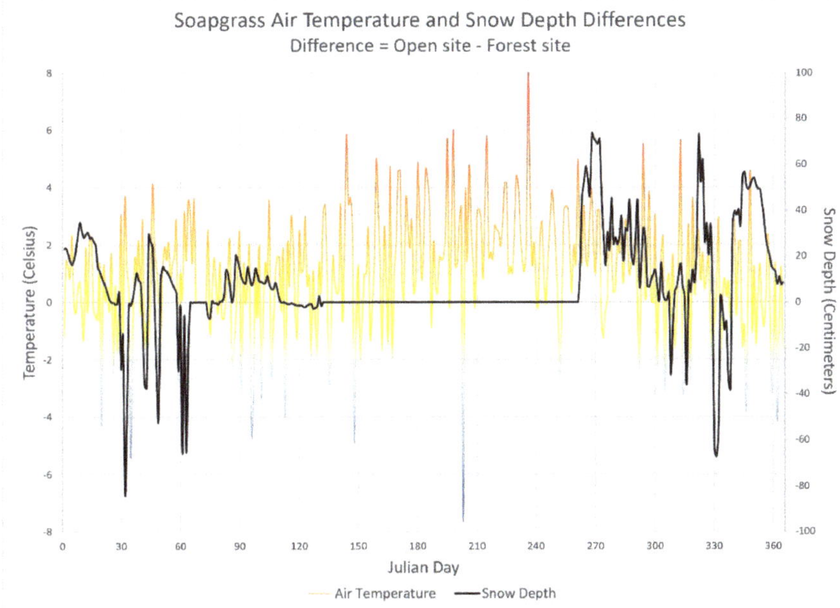

Figure 2. Daily air temperature (°C) and snow depth (cm) for both the forest and open sites at the Soapgrass station. A positive air temperature difference means the open site was warmer than the forest site. A positive depth difference means the open site had more snow than the forest site.

2.3. Original VELMA Air Soil Temperature (AST) Subroutine

For calculating spatially distributed soil temperature, VELMA accounts for soil moisture damping and the oscillatory effects of solar energy through a modified version of the Carslaw and Jaeger equation. This approach accounts for the seasonal solar energy variability through a time phase lag modification of observed air temperature combined with a temperature modification based on a soil depth attenuation. VELMA's subroutine, along with the equation previously presented by Carslaw and Jaeger (1959) (Equations (2) and (3)), does not account for spatial heterogeneity of solar energy reaching the ground due to topographic or object shading; instead, VELMA's input variables account for the shift in soil temperature due to daily air temperature, soil moisture and soil depth (Figure 3).

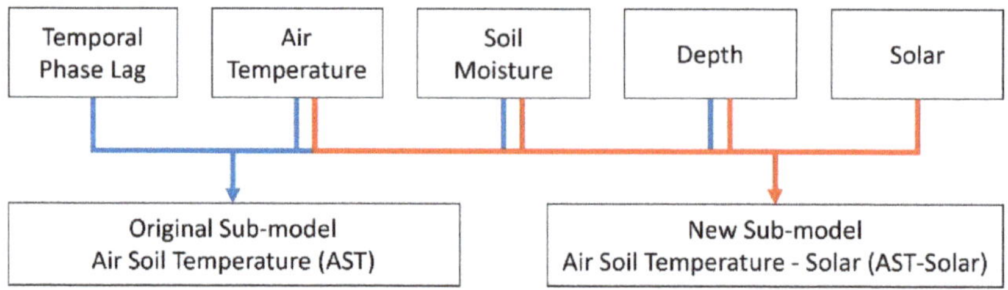

Figure 3. Combined AST and AST-Solar subroutines' schematic depicts the input data driving both models and displays the crucial data input difference between AST and AST-Solar with Temporal Phase Lag influencing only AST and Solar Energy influencing only AST-Solar.

For each layer, the AST subroutine calculates the soil temperature based on the thermal attenuation of daily air temperature. The input variables for the AST and AST-Solar models are mostly the same but might be utilized differently within each subroutine's equation setup, with the exceptions of the Air_{LAG} and $Reducer_{SOLAR}$ (Table 1).

Table 1. AST and AST-Solar subroutine input variables.

AST Variables	**Descriptions**
$Air_{AVETEMP}$	Fixed value of 8.2 (°C)
Air_{LAG}	Historic air temperature derived from the $Phase_{LAG}$.
LS_{DEPTH}	Soil column depth to center (mm) per layer of interest
$LTD_{ACCUMULATION}$	Summation of the thermal deltas per layer of interest
$Soil_{BELOW}$	Soil layer below the current layer being calculated
AST-Solar Variables	**Descriptions**
Air_{TEMP}	Daily average air temperature in the open site
$Soil_{AVE\_TEMP}$	Two-day running average (°C)
$Reducer_{SOLAR}$	Derived value from the input solar energy data
$Layers_{SM}$	Volume to volume soil moisture level
$SoilTemp_{(JDAY)}$	Current time steps soil temperature value
$SoilTemp_{(JDAY-1)}$	Prior time steps soil temperature value

The degree of attenuation is adjusted daily by the depth and soil moisture of each soil layer. G_{TEMP} is the resulting soil temperature due to: $Air_{AVETEMP}$ being the daily average air temperature (Table 1), Air_{LAG} (Equation (5) from a prior $Air_{AVETEMP}$ (Table 1) based on seasonal oscillation, $Soil_{DAMPING}$ (Equation (7) influencing the soil moisture based on seasonal oscillation, $Depth_{ATTENUATION}$ (Equation (9) based on soil depth and soil damping (Equation (7), $Phase_{LAG}$ (Equation (6) influenced by the $Air_{AVETEMP}$ (Table 1) based on seasonal oscillation driven by LS_{DEPTH} (depth to surface) (Table 1), $Soil_{DAMPING}$ (Equation (7), LTD (Equation (8) being the soil temperature accumulation at depth, and $Soil_{BELOW}$ (Table 1) influencing soil temperature from the lower soil layer:

$$G_{TEMP} = Air_{AVETEMP} + (Air_{LAG} - Air_{AVETEMP} - Soil_{DAMPING}) \times Depth_{ATTENUATION} \qquad (4)$$

$$Air_{LAG} = \text{Past Air Temperature at Julian Day's } Phase_{LAG} \qquad (5)$$

$$Phase_{LAG} = (LS_{DEPTH}/Soil_{DAMPING}) \times (365/2\pi) \qquad (6)$$

$$\text{Soil}_{\text{DAMPING}} = \sqrt{\frac{\text{LTD} \times 365}{\pi}} \qquad (7)$$

$$\text{LTD} = \text{LTD}_{\text{ACCUMULATION}} / \text{Soil}_{\text{BELOW}} \qquad (8)$$

$$\text{Depth}_{\text{ATTENUATION}} = e^{\wedge}(-\text{LS}_{\text{DEPTH}} / \text{Soil}_{\text{DAMPING}}) \qquad (9)$$

Each day, the air temperature is given as an input for each cell within VELMA's watershed framework. The $\text{Soil}_{\text{DAMPING}}$ (Equation (7) and LS_{DEPTH} (Table 1) variables are used to dampen the variations of soil temperature at larger depths. Any soil temperature shifts due to solar energy are incorporated via a proxy of two oscillatory equations driven by past air temperature (Equation (5) and soil moisture damping (Equation (7). VELMA AST subroutine's performances for open and forested landscapes are tested in the model testing section below.

2.4. New VELMA Air Soil Temperature-Solar (AST-Solar) Subroutine

The previously described soil temperature model does not utilize spatially explicit solar energy data, so we improved the model by adding the capacity to utilize spatially distributed ground-level solar energy to the current VELMA AST subroutine. This new model is called Air Soil Temperature-Solar (AST-Solar). The inclusion of solar energy within VELMA's original AST model was mainly accomplished through the addition of the new parameter $\text{Reducer}_{\text{SOLAR}}$ (Equation (12). Like a natural system, solar energy, via the variable $\text{Reducer}_{\text{SOLAR}}$ (Equation (12), only impacts the top soil layer, called $\text{NetSoil}_{\text{TEMP1}}$ (Equation (10). $\text{NetSoil}_{\text{TEMP1}}$ is defined as the following:

$$\text{NetSoil}_{\text{TEMP1}} = \text{Air}_{\text{TEMP}} \times \text{Reducer}_{\text{SOLAR}} \times \text{Damping}_{\text{SOIL}} \qquad (10)$$

where daily average air temperature (Air_{TEMP}; Table 1), $\text{Damping}_{\text{SOIL}}$ (Equation (11), and $\text{Reducer}_{\text{SOLAR}}$ (Equation (12) are multiplied together. AST soil moisture damping was included, yet simplified to only the inversion of each layer's fraction of volume to volume (v/v) soil moisture ($\text{Layers}_{\text{SM}}$; Table 1):

$$\text{Damping}_{\text{SOIL}} = 1 - \text{Layers}_{\text{SM}} \qquad (11)$$

Solar energy was built into the AST-Solar approach by accounting for the proportional relationship between each cell's solar energy to the landscape cell with the maximum solar energy. At each simulation timestep, the spatially distributed solar energy and the watershed's maximum solar energy at any location are both used to calculate the solar energy reduction called $\text{Reducer}_{\text{SOLAR}}$ (Equation (12). $\text{Reducer}_{\text{SOLAR}}$ represents the localized reduction in soil temperature due to shadowing in relation to the watershed's maximum ground-level solar energy. The reduction of solar energy is calculated as follows:

$$\text{Reducer}_{\text{SOLAR}} = 1 - \alpha \times (1 - (\text{Cell}_{\text{SOLAR}} / \text{Max}_{\text{SOLAR}})) \qquad (12)$$

where $\text{Cell}_{\text{SOLAR}}$ is each cell of interest within the VELMA framework, $\text{Max}_{\text{SOLAR}}$ is the landscape's maximum solar energy value amongst all landscape cells per time step, and α is a calibration factor. The calibration factor α is a fraction [0.0–1.0, where 0.0 is no solar energy and 1.0 is no change to the solar energy] that allows control over the influence of $\text{Reducer}_{\text{SOLAR}}$. But, to allow a direct and fair comparison of AST to AST-Solar, calibration factor α was not used in these tests.

VELMA utilizes four soil layers, and the thickness of each layer is customizable. VELMA soil temperature is not directly affected by solar energy, but rather through soil depth attenuation. For layers 2, 3 and 4, the soil temperature ($\text{NetSoil}_{\text{TEMPX}}$) is calculated using the 2-day running average temperature of the soil layer directly above, plus a reduction by $\text{Damping}_{\text{SOIL}}$ (Equation (11):

$$\text{NetSoil}_{\text{TEMPX}} = \text{Soil}_{\text{AVE\_TEMP}} \times \text{Damping}_{\text{SOIL}} \qquad (13)$$

$$\text{Soil}_{\text{AVE\_TEMP}} = (\text{SoilTemp}_{(\text{JDay})} + \text{SoilTemp}_{(\text{JDay}-1)})/2 \qquad (14)$$

where $SoilTemp_{(JDay)}$ (Table 1) is the current time steps soil temperature, and $SoilTemp_{(JDay-1)}$ (Table 1) is the prior time steps soil temperature. The soil moisture is applied as the $Damping_{SOIL}$ coefficient (Equation (11).

2.5. Subroutine Testing

We utilized data from EPA's O'CCMoN sites to test any change in accuracy and seasonal performance between the AST versus AST-Solar soil temperature subroutines. The O'CCMoN transect dataset provided observed driver data of air temperature, photosynthetic active radiation (PAR) as micromoles/meter^2/second (μmoles/m^2/s), and soil moisture as volume to volume at two soil layer depths [23]. EPA's observed O'CCMoN data helped to compare the AST versus AST-Solar models. Each O'CCMoN site also provided observed soil temperature at two depths. The soil temperature data were used to generate goodness-of-fit metrics against the simulated model results.

2.5.1. O'CCMoN Testing Sites

For model testing, the four following O'CCMoN locations were chosen: Cascade Head, Moose Mountain, Soapgrass, and Toad Creek. Each O'CCMoN location contains one forested site and one open clear-cut site (Figure 4).

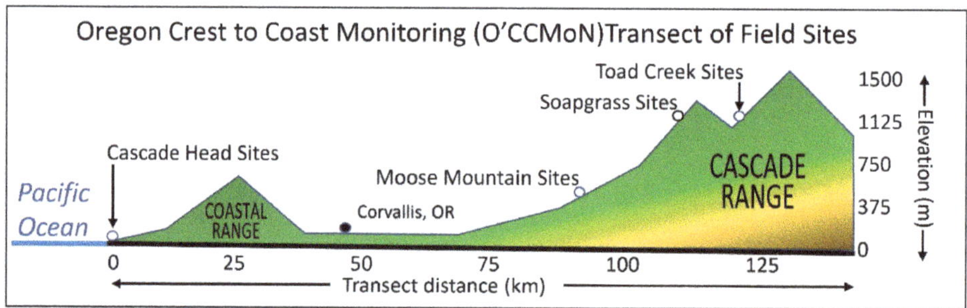

Figure 4. EPA Oregon Crest-to-Coast Environmental Monitoring (O'CCMoN) transit sites with environmental West to East trends and details for the sites used in this study. Figure adapted from the Crest to Coast Overview document [28].

Overall, these sites span a wide range of elevations and habitat diversities between the coast and the Cascade Mountain snow zone (Table 2).

The Cascade Head open site was installed outside the Cascade Head Experimental Forest and Scenic Research Area-Forestry Sciences Laboratory (EFSRA-FSL) in the managed landscape at an elevation of 157 m (Table 2). The Cascade Head forest site is located 190 m to the northeast in a predominantly Douglas-fir forest at an elevation of 190 m (Table 2). The Moose Mountain, Soapgrass, and Toad Creek sites are positioned on the western side of the Cascades Mountain Range at increasing elevations and experience moderate to extreme weather. The Moose Mountain open site was installed within a forest clear-cut at an elevation of 668 m with the forest site located 460 m to the northeast in a predominantly Douglas-fir forest at an elevation of 658 m (Table 2). The Soapgrass open site was installed within a forest clear-cut at an elevation of 1298 m with the forest site located 1190 m to the northeast in a predominantly Douglas-fir forest also at an elevation of 1190 m (Table 2). The Toad Creek open site was installed within a forest clear-cut at an elevation of 1202 m with the forest site located 471 m to the east in a predominantly Douglas-fir forest at an elevation of 1198 m (Table 2).

The soil temperature probes were all installed in the same manner at all EPA O'CCMoN locations for both the open site and forested site. In each location, two soil temperature sensors were installed at a depth of 15 cm and 30 mm, respectively, from the mineral soil surface, i.e., just below the O-horizon [23]. The testing of the AST and AST-Solar subroutines did not involve any data collection, but rather leveraged the data collected through the EPA O'CCMoN project. These data and the details of field work can be found in the documents at the data repository [23].

Table 2. O'CCMoN Open and Forest Site Characteristics.

Site Name	Elevation (m)	Vegetative State	Annual Rainfall (cm)	Tree Height (m)	Soil Parent Material
Cascade Head: Open (CHO)	157	Lawn		-	
Cascade Head: Forest (CH14)	190	Alder Douglas-fir Sitka Spruce	200–250	50–60	Marine Sediment
Moose Mountain: Open (MMO)	668	Clear-cut	150–180	-	Volcanic
Moose Mountain: Forest (MMF)	658	Douglas-fir		50–60	
Soapgrass: Open (SGO)	1298	Clear-cut	180–200	-	Volcanic
Soapgrass: Forest (SGF)	1190	Douglas-fir		60–70	
Toad Creek: Open (TCO)	1202	Clear-cut	180–200	-	Volcanic
Toad Creek: Forest (TCF)	1198	Douglas-fir		50–60	

Note: All information in this table was obtained from the O'CCMoN dataset documentation, except the annual rainfall which was obtained through the PRISM 1981–2010 annual rainfall normals [29].

2.5.2. AST versus AST-Solar Subroutine Setup

The AST and AST-Solar subroutines were both ran from 1 January 2005 through 31 December 2005 at a daily time step. For each site, the same O'CCMoN observed air temperature and soil moisture data were used as the data drivers for both subroutines [23]. The O'CCMoN data are measured in 30-min intervals, yet the VELMA model functions at a daily time step. To match the VELMA temporal grain, all observed O'CCMoN data were averaged to a 24-h period.

The VELMA spatial framework, per cell, contains four voxel layers; therefore, the AST and AST-Solar subroutines function under this spatial framework. Yet, the O'CCMoN dataset contains soil temperature probe data at only two depths. The AST and AST-Solar soil moisture probe depth variables for layer 1 and 2 were set to match the sensor depths of 15 cm and 30 cm [23]. Since the O'CCMoN data sites contained only two soil moisture probe depths for the sites selected, the AST and AST-Solar voxel layer three and four soil temperature results could not be evaluated and were excluded.

The AST-Solar model utilized the additional solar energy driver data. For the AST-Solar open site simulations, the $Cell_{SOLAR}$ and Max_{SOLAR} variables utilized open site solar energy data. For the AST-Solar forest site simulations, the $Cell_{SOLAR}$ variable utilized the forest site solar data, while the Max_{SOLAR} variable was calculated using the open site solar energy data.

The variable $Reducer_{SOLAR}$ is utilized as a fractional variable scaled from zero to one (Equation (12). This setup allows any solar energy units to be implemented through this method. $Cell_{SOLAR}$ being the solar energy per location and Max_{SOLAR} representing the location with the most solar energy means for an open site, the $Cell_{SOLAR}$ and Max_{SOLAR} values will similar if not the same. In this scenario, $Reducer_{SOLAR}$ will cause minimal to no reduction to the soil temperature. Conversely, forest site $Cell_{SOLAR}$ and Max_{SOLAR} values will be quite different. In this scenario, $Reducer_{SOLAR}$ will cause a reduction in the soil temperature.

3. Results

Model results for each of the O'CCMoN sites are summarized in Table 3. Overall, the inclusion of spatially distributed solar energy improved the simulated solar temperature results. Both open and forested sites exhibit gains in accuracy, though the inclusion of spatially distributed solar energy was most beneficial for the forest sites. Below, all data are distinguished using the following attributes: site location, open versus forest environment, and soil layer. O'CCMoN sites with open versus forest locations are listed with abbreviations under "Sites" in Table 3. Soil Layer 1 and Soil Layer 2 are referred to as SL1 and SL2, respectively.

Table 3. VELMA-AST and VELMA-AST3 O'CCMoN results.

O'CCMoN Location	Sites	Soil Layer 1		Soil Layer 2	
		AST (r^2)	AST3 (r^2)	AST (r^2)	AST3 (r^2)
Cascade Head	Open Site (CHO)	0.83	0.76	0.71	0.95
	Forest Site (CH14)	0.74	0.87	0.71	0.94
Moose Mountain	Open Site (MMO)	0.81	0.92	0.67	0.93
	Forest Site (MMF)	0.89	0.93	0.70	0.94
Soapgrass	Open Site (SGO)	0.80	0.85	0.69	0.90
	Forest Site (SGF)	0.69	0.92	0.57	0.89
Toad Creek	Open Site (TCO)	0.82	0.83	0.72	0.92
	Forest Site (TCF)	0.83	0.90	0.64	0.89

The performance of the AST-Solar model at the Soapgrass site increased for soil layers 1 and 2 at both the open and forest sites compared to the AST model, but especially for soil layer 2 (Table 3). Specifically, the SGO-SL1 performance increased from a r^2 of 0.80 to 0.85 (Table 3; Figure 5A), while the SGO-SL2 performance increased from a r^2 of 0.69 to 0.90 (Table 3; Figure 5C). The SGF-SL1 performance increased from a r^2 of 0.69 to 0.92 (Table 3; Figure 5B), while SGF-SL2 performance increased from a r^2 of 0.57 to 0.89 (Table 3; Figure 5D).

The results among all sites are unique for each site, though the seasonal pattern and increased performance are similar over the year. Since the patterns are similar for each of the different sites, only the Soapgrass site results are graphically represented. Due to the 100-day Phase$_{LAG}$ (spin-up) requirement of the AST model, only Julian days 101 through 365 are graphically represented.

The performance at Cascade Head (CH) increased for the AST-Solar subroutine compared with the AST subroutine for soil layers 1 and 2 for the forest site, but the performance only improved soil layer 2 of the open site. The open site soil layer 1 was the only simulation that exhibited a decrease in simulated versus observed agreement by decreasing from a r^2 of 0.83 to 0.77. In contrast, the CHO-SL2 performance increased from a r^2 of 0.71 to 0.95. All CH14 simulations improved. The CH14-SL1 performance increased from a r^2 of 0.74 to 0.87, while the CH14-SL2 performance increased from a r^2 of 0.71 to 0.94.

The performance of the AST-Solar subroutine compared with the AST subroutine at the Moose Mountain (MMO) increased for soil layers 1 and 2 at both the open and forest sites, particularly for soil layer 2. The MMO-SL1 performance increased from a r^2 of 0.81 to 0.92 (Table 3). The MMO-SL2 performance increased from a r^2 of 0.67 to 0.93 (Table 3). The MMF-SL1 performance increased from a r^2 of 0.89 to 0.93 (Table 3). The MMF-SL2 performance increased from a r^2 of 0.70 to 0.94 (Table 3).

The performance of the AST-Solar subroutine over the AST subroutine at the Toad Creek increased for soil layers 1 and 2 at both the open and forest sites. The TCO-SL1 performance increased from a r^2 of 0.82 to 0.83, while the TCO-SL2 performance increased from a r^2 of 0.73 to 0.92 (Table 3). The TCF-SL1 performance increased from a r^2 of 0.83 to 0.90, while the SGF-SL2 performance increased from a r^2 of 0.64 to 0.89.

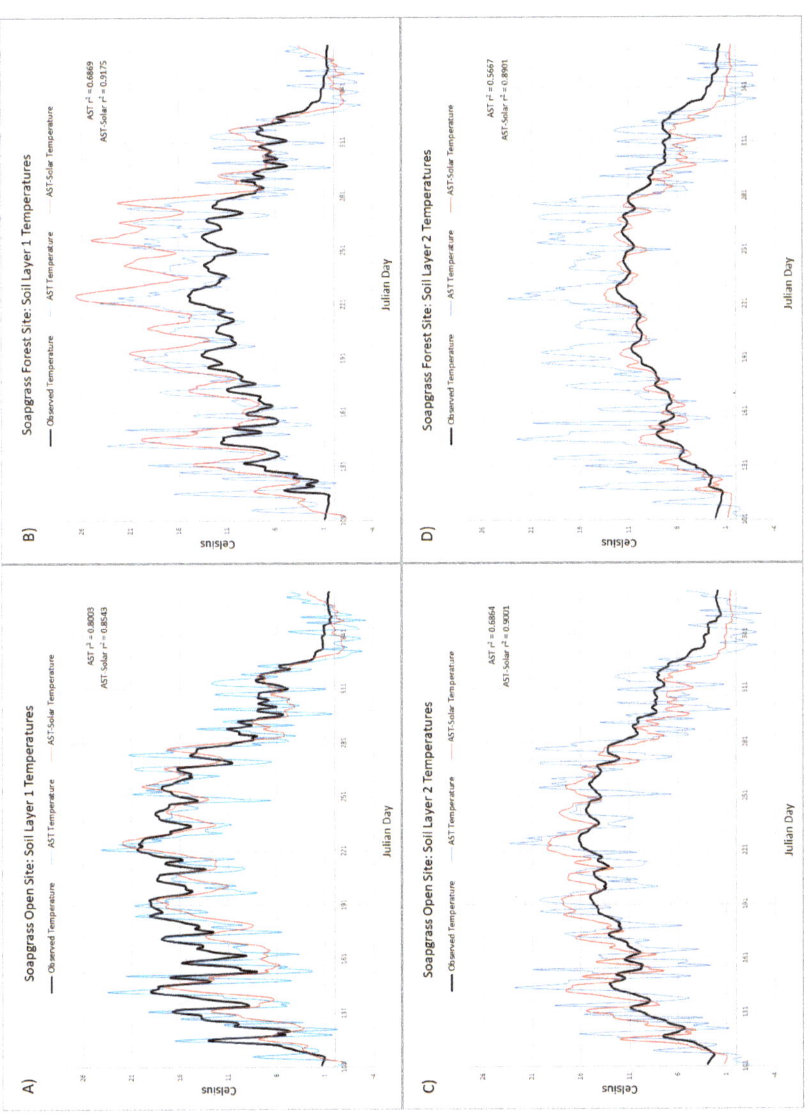

Figure 5. All panels compare observed data for the year 2005, AST model results, and AST-Solar model results. (Panel **A**) Soapgrass Open Site Soil Layer 1 temperature results. (Panel **B**) Soapgrass Forest Site Soil Layer 1 temperature results. (Panel **C**) Soapgrass Open Site Soil Layer 2 temperature results. (Panel **D**) Soapgrass Forest Site Soil Layer 2 temperature results. Simulation r^2 values are calculated using Julian days 101–365. Ignoring the first 100 days prevented the AST model from being penalized for its 100-day phase lag, which only occurs during the first year of simulation.

4. Discussion

VELMA's original soil temperature model functioned well without the inclusion of spatially distributed solar energy (see Table 3; Figure 5), yet the inclusion of spatially distributed solar energy can provide significant improvements to simulated ecological processes driven by solar energy. The inclusion of local ground-level solar energy data improved VELMA's AST subroutine simulations of soil temperature from more than one perspective. First, the observed versus modeled comparisons for all sites improved, with one exception. The only exception was the CHO site, which was a landscape anomaly amongst the sites due to the station existing within a regularly maintained grass lawn at the headquarters of the research area (i.e., Cascade Head EFSRA-FSL). The TCO site received the smallest soil temperature modeling improvement with SL1 r^2 increasing from 0.82 to 0.83, yet the TCF site showed a significant SL2 improvement with a r^2 increase of 0.73 to 0.92. The largest single layer improvement was observed at the SGF site with the SL1 r^2 increasing from 0.69 to 0.92 and SGF-SL2 r^2 increasing from 0.57 to 0.89. It is worth reiterating that though the AST-Solar subroutine has a calibration parameter, no calibration was applied when simulating any of the sites to ensure the AST to AST-Solar estimates of soil temperature were a fair comparison of the subroutine performance.

Beyond the r^2 goodness-of-fit metrics, the daily variability in the modeled AST-Solar soil temperature data was reduced. This can be seen in all the graphs presented above by comparing how the AST subroutine demonstrated a repeated overestimation and underestimation of soil temperature compared to the observed data. However, the AST-Solar subroutine's noise was greatly reduced. This noise pattern in AST was due to the significant influence from the daily average air temperature driver data. Therefore, the static equations that provided oscillatory proxies for solar energy did not fully parallel the environmental phenomena of solar energy. In part, this disparity explains the resulting noisy estimations of soil temperature when solar energy was not directly included in the subroutine.

VELMA's AST soil temperature subroutine required a 100-day soil temperature simulation spin-up period. This time lag allowed for a sufficient temporal delay in the driver data utilized by Equation (6) (note that this time lag is only required for the first year of multi-year AST simulations). The purpose of the lag time was to account for the seasonal weather influence on the soil column. For the new AST-solar subroutine, this lag was no longer required due to the addition of localized daily solar energy data interacting with the existing soil moisture within layer 1.

Both the AST and AST-Solar subroutines do not model the insulation effects of snow pack [4]. Further work could include snow depth and its insulative effects on soil temperature. This would improve the soil temperature estimates in the winter, due the insulation of heat from the snow pack preventing the soil temperature from getting colder or even freezing. For the AST-Solar model, this may further improve performance with observed data (Figure 5, panels A–D) for days where snow pack persisted at Soapgrass (Figure 2) as well as goodness-of-fit metrics for the other O'CCMoN monitored sites. Similarly, further improvements in AST-Solar soil temperature estimates may be possible by including VELMA predictions of surface detritus (dead leaves and wood) and ground-level leaf biomass (proxy for leaf area index) that contribute to near-surface shading of mineral soil surfaces in open and forest sites.

The subroutine performance improvements reported here are due to the influence of local solar energy, which alters the resulting soil temperature. The prior soil temperature model predominantly utilized average air temperature as a proxy for energy, which is commonly done in watershed models. Though VELMA is a spatially distributed model, the default weather model is driven by single site location climate data. This setup resulted in homogeneous average air temperature across the simulated watershed. This is true for all forest cells and bare open prairie or forest clear-cut cells alike. By including local solar energy representation, the subsequent modeling of soil temperature was enhanced due to the improved model representation of the real world. This mainly was accomplished through the inclusion of the environmental variable solar energy that causes direct and significant influence on the phenomena soil temperature.

5. Conclusions

Watershed models are widely used to simulate the effects of land use change on the environment and the quantity and quality of hydrologic components throughout a watershed. In this paper, we demonstrated that local solar energy information improved soil temperature modeling estimates simulated by a soil temperature subroutine within a larger ecohydrological watershed model. These models were compared with observed data for soil temperatures at two depths within both open and forested environments among four observed data sites (i.e., EPA's O'CCMoN transect data) [23].

Overall, by including explicit information regarding the spatial distribution of solar energy across a landscape, watershed models can better capture the spatiotemporal variations of soil temperature in both forested and open sites. Therefore, researchers that utilize spatially distributed or semi-distributed mechanistic watershed models should consider incorporating spatially explicit solar energy models (e.g., Penumbra [30,31]) or other spatially heterogeneous descriptions of ground-level solar energy to better represent energy exchange at the surface. This is especially true when modeling discrete landcover types such as forested, open, water, and agricultural cover and when modeling the impacts of riparian shading on soil temperature and stream temperature, as well as the effect of solar energy on fish habitat.

Finally, while we presented improvements of a soil temperature subroutine within an ecohydrological model, other subroutines can also benefit by the inclusion of spatiotemporal representations of ground-level solar energy. The integration of local solar energy information with watershed models and all their subroutines could potentially benefit several processes, such as snow melt, water temperature, and plant growth via photosynthesis. Integration of spatially explicit ground-level solar energy models with environmental models can provide dynamic feedbacks between other environmental processes, such as tree growth and shade. As tree growth is simulated within watershed models, their heights could be transferred back to the ground-level solar energy model in a dynamic mechanism, which then would alter the amount of solar energy that is intercepted by the canopy and does not reach the ground. This dynamic integrated modeling approach could be extremely beneficial for looking at the long-term effects of planting riparian buffers and determining the duration required for stream temperatures to be reduced by some threshold. Because solar energy is amongst the most impactful environmental variables in natural and managed ecosystems, further investigations would greatly benefit from the application of watershed-scale models that are dynamically coupled with spatially-explicit solar energy models.

Author Contributions: J.J.H., B.L.B., R.B.M., R.E.K. and J.J.G. designed the project and modeling approach; J.J.H. and B.L.B. performed, and R.B.M., R.E.K., A.F.B., K.S.D., and P.P.P. assisted with, the model set up and simulations; R.S.W performed all field data curation; J.J.H., B.L.B., R.E.K., R.B.M., J.J.G. and R.S.W. analyzed the results and developed the manuscript's structure; all authors contributed to writing the paper.

Funding: This research received no external funding.

References

1. Mayer, T.D. Controls of Summer Stream Temperature in the Pacific Northwest. *J. Hydrol.* **2012**, *475*, 323–335. [CrossRef]
2. Reich, P.B.; Luo, Y.; Bradford, J.B.; Poorter, H.; Perry, C.H.; Oleksyn, J. Temperature drives global patterns in forest biomass distribution in leaves, stems, and roots. *Proc. Natl. Acad. Sci. USA* **2014**, *111*, 13721–13726. [CrossRef] [PubMed]
3. Mellander, P.-E.; Stähli, M.; Gustafsson, D.; Bishop, K. Modelling the effect of low soil temperatures on transpiration by Scots pine. *Hydrol. Process.* **2006**, *20*, 1929–1944. [CrossRef]

4. Stieglitz, M.; Ducharne, A.; Koster, R.; Suarez, M. The impact of detailed snow physics on the simulation of snow cover and subsurface thermodynamics at continental scales. *J. Hydrometeorol.* **2001**, *2*, 228–242. [CrossRef]

5. Abdelnour, A.; Stieglitz, M.; Pan, F.; Mckane, R. Catchment hydrological responses to forest harvest amount and spatial pattern. *Water Resour. Res.* **2011**, *47*, 1995–2021. [CrossRef]

6. Neitsch, S.L.; Arnold, J.G.; Kiniry, J.R.; Williams, J.R.; King, K.W. *Soil and Water Assessment Tool Theoretical Documentation Version 2009*; Texas Water Resources Institute, Texas A&M University System: College Station, TX, USA, 2011; pp. 32–38.

7. Tague, C.L.; Band, L.E. RHESSys: Regional hydro-ecologic simulation system—An object-oriented approach to spatially distributed modeling of carbon, water, and nutrient cycling. *Earth Interact.* **2004**, *8*, 1–42. [CrossRef]

8. Bicknell, B.R.; Imhoff, J.C.; Kittle, J.L.; Donigian, A.S.; Johanson, R.C. *Hydrological Simulation Program—Fortran, User's Manual for Version 11*; U.S. Environmental Protection Agency, National Exposure Research Laboratory: Athens, GA, USA, 1997; p. 755.

9. National Weather and Climate Center (NWCC). *Snow Telemetry (SNOTEL) Data Collection Network*; Natural Resources Conservation Service (NRCS): Portland, OR, USA, 2016.

10. Reynolds, C.J., III. Aviation Weather Information Dissemination System. U.S. Patent 4,521,857, 4 June 1985.

11. Foundation, N.S. LTER Network. Available online: https://lternet.edu/ (accessed on 26 September 2018).

12. OSU Research Forests. Available online: http://cf.forestry.oregonstate.edu/ (accessed on 26 September 2018).

13. Harvard Forest. Available online: https://harvardforest.fas.harvard.edu/ (accessed on 26 September 2018).

14. Fuka, D.R.; Walter, M.T.; MacAlister, C.; Degaetano, A.T.; Steenhuis, T.S.; Easton, Z.M. Using the Climate Forecast System Reanalysis as weather input data for watershed models. *Hydrol. Process.* **2013**, *28*, 5613–5623. [CrossRef]

15. Daly, C.; Halbleib, M.; Smith, J.I.; Gibson, W.P.; Doggett, M.K.; Taylor, G.H.; Curtis, J.; Pasteris, P.P. Physiographically sensitive mapping of climatological temperature and precipitation across the conterminous United States. *Int. J. Climatol.* **2008**, *28*, 2031–2064. [CrossRef]

16. Thornton, P.E.; Running, S.W.; White, M.A. Generating surfaces of daily meteorological variables over large regions of complex terrain. *J. Hydrol.* **1997**, *190*, 214–251. [CrossRef]

17. Staff, S.S.D. Soil survey manual. In *USDA Handbook 18*; Ditzler, C., Scheffe, K., Monger, H.C., Eds.; Government Printing Office: Washington, DC, USA, 2017.

18. Li, G.; Jackson, C.R.; Kraseski, K.A. Modeled riparian stream shading: Agreement with field measurements and sensitivity to riparian conditions. *J. Hydrol.* **2012**, *428–429*, 142–151. [CrossRef]

19. Boyd, M.; Kasper, B. *Analytical Methods for Dynamic Open Channel Heat and Mass Transfer: Methodology for the Heat Source Model Version 7.0*; Watershed Sciences Inc.: Portland, OR, USA, 2003.

20. Seidl, R.; Rammer, W.; Scheller, R.M.; Spies, T.A. An individual-based process model to simulate landscape-scale forest ecosystem dynamics. *Ecol. Model.* **2012**, *231*, 87–100. [CrossRef]

21. Chiacchio, M.; Solmon, F.; Giorgi, F.; Stackhouse, P.; Wild, M. Evaluation of the radiation budget with a regional climate model over Europe and inspection of dimming and brightening. *J. Geophys. Res. Atmos.* **2015**, *120*, 1951–1971. [CrossRef]

22. Alexandri, G.; Georgoulias, A.K.; Zanis, P.; Katragkou, E.; Tsikerdekis, A.; Kourtidis, K.; Meleti, C. On the ability of RegCM4 regional climate model to simulate surface solar radiation patterns over Europe: An assessment using satellite-based observations. *Atmos. Chem. Phys.* **2015**, *15*, 13195–13216. [CrossRef]

23. Waschmann, R.S.; U.S. Environmental Protection Agency—Western Ecology Division. *Oregon Crest-to-Coast Climate Observations*; NOAA National Centers for Environmental Information: Corvallis, OR, USA, 2014.

24. Ficklin, D.L.; Luo, Y.; Stewart, I.T.; Maurer, E.P. DeveloFpment and application of a hydroclimatological stream temperature model within the Soil and Water Assessment Tool. *Water Resour. Res.* **2012**, *48*. [CrossRef]

25. Bokhorst, S.; Huiskes, A.; Aerts, R.; Convey, P.; Cooper, E.J.; Dalen, L.; Erschbamer, B.; Gudmundsson, J.; Hofgaard, A.; Hollister, R.D.; et al. Variable temperature effects of Open Top Chambers at polar and alpine sites explained by irradiance and snow depth. *Glob. Chang. Biol.* **2012**, *19*, 64–74. [CrossRef] [PubMed]

26. McKane, R.B.; Brookes, A.; Djang, K.; Stieglitz, M.; Abdelnour, A.G.; Pan, F.; Halama, J.J.; Pettus, P.B.; Phillips, D.L. *VELMA Version 2.0: User Manual and Technical Documentation*; U.S. Environmental Protection Agency—Western Ecology Division: Corvallis, OR, USA, 2014; p. 60.

27. Carslaw, H.S.; Jaeger, J.C. *Conduction of Heat in Solids*, 2nd ed.; Clarendon Press: Oxford, UK, 1959.

28. Waschmann, R.S. *EPA Oregon Crest to Coast Overview, Figure 2*; NOAA National Centers for Environmental Information: Corvallis, OR, USA, 2018; p. 16.

29. Daly, C. 30-Year Normals. Available online: http://www.prism.oregonstate.edu/normals/ (accessed on 26 September 2018).

30. Halama, J.J. Penumbra: A Spatiotemporal Shade-Irradiance Analysis Tool with External Model Integration for Landscape Assessment, Habitat Enhancement, and Water Quality Improvement. Ph.D. Thesis, Oregon State University, Corvallis, OR, USA, 2017.

31. Halama, J.J.; Kennedy, R.E.; Graham, J.J.; McKane, R.B.; Barnhart, B.L.; Djang, K.; Pettus, P.B.; Brookes, A.; Wingo, P.C. Penumbra: A spatially-distributed shade percent and incident ground-level irradiance model for simulating landscape scale solar energy. *PLoS ONE* **2018**, in press.

Article

Issues of Meander Development: Land Degradation or Ecological Value? The Example of the Sajó River, Hungary

László Bertalan [1,*], Tibor József Novák [2], Zoltán Németh [3], Jesús Rodrigo-Comino [4,5], Ádám Kertész [6] and Szilárd Szabó [1,6]

[1] Department of Physical Geography and Geoinformatics, University of Debrecen, Egyetem tér 1, H-4032 Debrecen, Hungary; szabo.szilard@science.unideb.hu

[2] Department of Landscape Protection and Environmental Geography, University of Debrecen, Egyetem tér 1, H-4032 Debrecen, Hungary; novak.tibor@science.unideb.hu

[3] Department of Evolutionary Zoology and Human Biology, University of Debrecen, Egyetem tér 1, H-4032 Debrecen, Hungary; nemethzoltan@science.unideb.hu

[4] Department of Geography, Instituto de Geomorfología y Suelos, Málaga University, Campus of Teatinos, 29071 Málaga, Spain; rodrigo-comino@uma.es

[5] Department of Physical Geography, Trier University, 54286 Trier, Germany

[6] Geographical Institute, Research Center for Astronomy and Earth Sciences, Hungarian Academy of Sciences, Budaörsi út 45, H-1112 Budapest, Hungary; kertesza@helka.iif.hu

[*] Correspondence: bertalan@science.unideb.hu; Tel.: +36-52-512-900 (ext. 22201)

Received: 29 September 2018; Accepted: 6 November 2018; Published: 9 November 2018

Abstract: The extensive destruction of arable lands by the process of lateral bank erosion is a major issue for the alluvial meandering type of rivers all around the world. Nowadays, land managers, stakeholders, and scientists are discussing how this process affects the surrounding landscapes. Usually, due to a land mismanagement of agroforestry activities or urbanization plans, river regulations are designed to reduce anthropogenic impacts such as bank erosion, but many of these regulations resulted in a degradation of habitat diversity. Regardless, there is a lack of information about the possible positive effects of meandering from the ecological point of view. Therefore, the main aim of this study was to investigate a 2.12 km long meandering sub-reach of Sajó River, Hungary, in order to evaluate whether the process of meander development can be evaluated as a land degradation processes or whether it can enhance ecological conservation and sustainability. To achieve this goal, an archive of aerial imagery and UAV (Unmanned Aerial Vehicle)-surveys was used to provide a consistent database for a landscape metrics-based analysis to reveal changes in landscape ecological dynamics. Moreover, an ornithological survey was also carried out to assess the composition and diversity of the avifauna. The forest cover was developed in a remarkable pattern, finding a linear relationship between its rate and channel sinuosity. An increase in forest areas did not enhance the rate of landscape diversity since only its distribution became more compact. Eroding riverbanks provided important nesting sites for colonies of protected and regionally declining migratory bird species such as the sand martin. We revealed that almost 70 years were enough to gain a new habitat system along the river as the linear channel formed to a meandering and more natural state.

Keywords: bank erosion; landscape metrics; diversity; Sajó River; UAV

1. Introduction

Alluvial rivers represent dynamic landforms of the watersheds all over the world [1]. Under minimal regulation, these rivers can generate diverse landscapes by shifting back and forth along their floodplain [2,3]. In meandering rivers, the channel flow undercuts the outside banks resulting in seepage outflow or mass failure processes i.e., slab failures, rotational slides and sloughings or slump blocks eroding into the water body [4,5]. These materials can form point bar accretion at the inside banks downstream due to the lower current velocity at the inner side, as opposed to the outer side of the bends [6]. The cross-sections of meandering riverbeds are complex showing asymmetric bathymetric and flow patterns, which is reflected in habitat diversity as well [7]. Point bars are specific fluvial geomorphic features inside a meandering river bend [8]. They are deposited from finer sediments by the low energy parts of the river along the inside bank driven by recirculation zones [9]. The ridge and swale topography produce undulating bar surfaces [10] and the long-inundated periods of the depressions allow pioneer bush and tree species colonizing the fine-grained surfaces. Intensive channel migration maintains ideal conditions for vegetation succession processes [11], since the bend migrates forward through lateral aggradation, the colonization process increases [12]. Generally, the maxima of lateral bank erosion rates are concentrated downstream of the point bar symmetry-axis [11] resulting elongated, skewed or compound meander bends [13].

River channels and their surrounding floodplains enhance landscape evolution and the diversification of environments [14]. Quasi-natural rivers without extensive channelization, bank protection and embankments, are susceptible to rapid changes in their hydro-geomorphological features. One of the main reasons for this is that vegetation colonizing riparian zones are affected by frequent flood-disturbances or even severe droughts, therefore, they have adapted to variable water levels. These plants have a major influence on the initiation of geomorphic and hydraulic processes of the channel and floodplain as they regularly stabilize organic matter from sediment fluxes [15]. The majority of these species need sunlight for growth that is not available under dense canopy levels; therefore, they colonize freshly formed open spaces, and the resulting patchy vegetation mosaics well-represent the habitat diversity established through flooding, scouring and sedimentation [16]. Furthermore, the resistance properties of the vegetation on channel flow is recognized as a potential factor in the mitigation of land degradation processes [17,18]. River corridors, as linear features of fluvial landscapes, are such integrated ecological systems that connect identical landscape elements. Due to the dynamic interactions within climatic factors, catchment geology, relief, inundation and nutrients, the ecological turnover can be outstanding in these features [12]. The aquatic and terrestrial vegetation patches provide heterogeneous and diverse habitat for fish, water birds and macroinvertebrates [19]. The initial colonizers maintain their stands, preventing further recruitment of vegetation beneath their canopy level [20,21]. Furthermore, vegetation patches substantially block and divert river flow around their canopy, and flow velocities decrease along the vegetation patch and increase in the surrounding channel area; thus, they determine the future occupancy patterns of trees along the point bars [22]. The succession of even ephemerally emerging bar surfaces provides suitable foraging and nesting sites for water birds. Moreover, due to their isolated nature, these sites are often free from human disturbance as well [19,23].

Lateral bank erosion of meandering rivers is responsible for extensive destruction of arable lands and usually threatens human environments [24,25]. These phenomena, accompanied by riparian deforestation and serious flood inundation, provide the necessity for the protection of most European river networks [26–28]. Regulated river channels have less cross-sectional diversity from both geomorphological and ecological point of view. In the past years, several studies revealed that channelization and bank protection works are responsible for an extensive, global-scale habitat and river ecosystem degradation [29–32]. It is also proven that river regulation impacted fish and macroinvertebrates negatively due to the degradation of habitat heterogeneity [33]. Recent studies found 're-meandering' as an effective practice for the restoration of habitat diversification at short- and medium-terms [34], since lateral erosion and the channel migration maintain sediment supply for vegetation colonization [35]. However, long-term ecological consequences of re-meandering have not been well-studied yet [36]. Only a few studies found positive trends in diversification but many restoration projects were not able to demonstrate significant differences in biodiversity between the before and after periods of the restoration project over short (5 to 10 years) time periods [9,37,38].

In Hungary, the majority of rivers had been regulated and channelized, starting in the 19th century [39–42]. However, few medium-size rivers of the Tisza River Basin such as the Sajó River remained in a quasi-natural state since economic issues had prevented the total channelization [43]. Along with several reaches of Sajó River, extensive lateral erosion threatens the agricultural areas. However, there is a lack of information about this vital issue, which could be included as key information for land management plans. Thus, the main aims of this research were (i) to investigate the geomorphological development and effects of bank erosion along a meandering sub-reach of Sajó River; and, (ii) to assess the possible impacts on ecological diversity in case there will not be further interventions on channel morphology. To achieve these goals, we performed a GIS (Geographic Information System)-analysis of three consecutive meandering bends over 10 periods between 1952 and 2017 based on archive aerial imagery and UAV-surveys. Moreover, an ornithological survey was also carried out to assess the composition and diversity of the avifauna.

2. Materials and Methods

2.1. Study Area

Sajó River (or Slaná in Slovakia) is a transboundary river of Slovakia and Hungary having a total length of 229 km (124 km in Hungary). It is the main tributary of Tisza River before it reaches the Great Hungarian Plain. The river catchment is situated at the Eastern-Carpathians with a total area of 5545 km^2. The stream gradient is much higher in Slovakian territory, then the river becomes alluvial downstream from the Hungarian state border. The mean discharge of the river is around 24 m^3/s. The total average of suspended sediment load varies between 828,000 m^3/y and 1,927,000 m^3/y [44]. The Hungarian reach of Sajó River is mainly the alluvial meandering-type with a total sinuosity of 1.78.

Although extensive river regulation plans had been established in the early 20th century to facilitate shipping towards the Carpathians, eventually, most of the works had to be cancelled due to the economic issues associated with World Wars [45]. Only a few minor river management works had been carried out mainly around industrial areas. Most of these are artificial cutoffs, small groynes, and another bank protection. Even though their spatial distribution is broad (58.3%) [43,46] along the Hungarian reach of Sajó River, this is one of the least regulated rivers in the country. The abovementioned geographical settings and the low rate of human intervention lead to the situation of lateral bank erosion that can reach the 4–7 m/y rate in several sub-reaches [46,47].

The Hungarian National Ecological Network (Phase 2) had been established between 1999–2001 according to the legislation of the Pan-European Ecological Network (PEEN). The components of the network provide the following landscape element categories: well-known core-areas, ecological corridors, buffer zones and restoration areas [48]. Core areas provide the main habitats and genetic reserves while the strip-like ecological corridors serve as continuous habitats or chains linking the smaller and larger habitat patches together [49]. This study focuses on a selected 2.12 km long sub-reach of the Sajó River consisting of three consecutive river bends near the town of Nagycsécs, Hungary (Figure 1). The further calculations were performed on this 68.4 ha large rectangular area around the river channel. This sub-reach of Sajó River can be considered as a free-forming meandering type located between major river engineering works both upstream and downstream. The study area as part of the Sajó River floodplain belongs to the PEEN category of ecological corridors between two main Natura2000 areas.

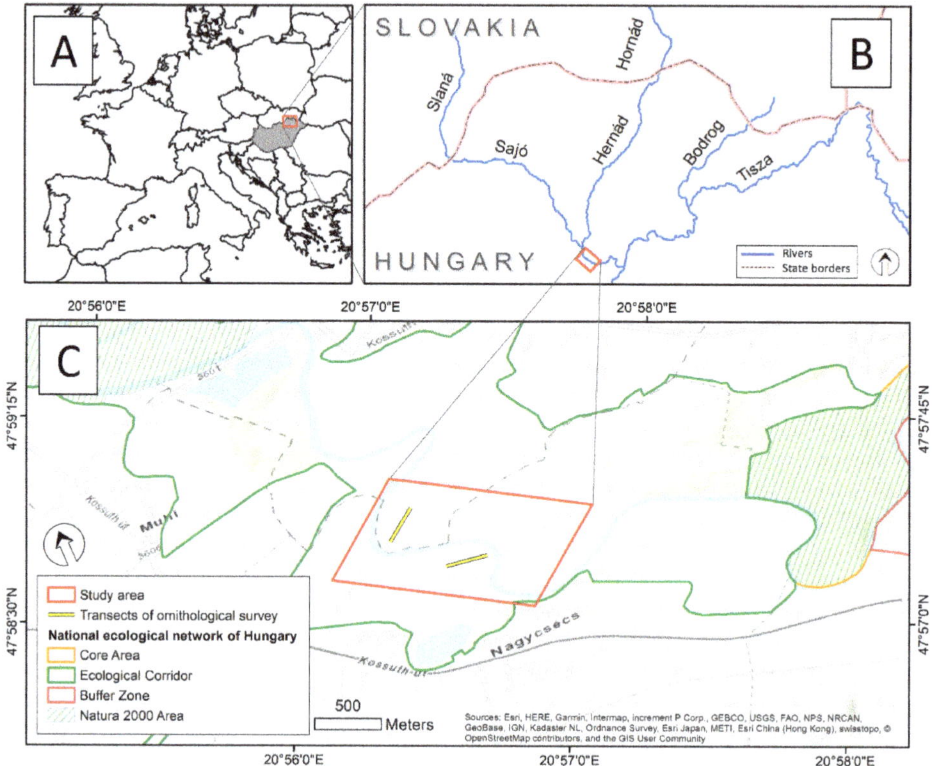

Figure 1. Overview of the study area. (**A**) Location of the study area in Europe; (**B**) Location of the main rivers in the region of study area; (**C**) Detailed overview of the selected sub-reach of Sajó River.

We observed that the land mosaic of the study area was represented by a patch-complex consisting of bare surfaces, perennial grasslands, forests and bushes, the river channel, arable lands, and settlements. Bare plots, grasslands, and arboreal vegetation were considered as patches connected to each other by their own successive development controlled by the river channel changes and flood dynamic. On the other, arable lands and settlements were considered as mainly human-controlled elements of this structure. Bare surfaces meant patches not completely in a lack of vegetation, but with a low cover. Plant species of these sites are mainly disturbance-tolerant species such as *Chenopodium album, Chenopodium ambrosioides, Bidens tripartitus*, at their wet riverside edges *Polygonum hydropier, Polygonum minus, Polygonum mite, Rorippa amphibia, Rumex crispus*, and *Rumex obtusifolius*. Perennial grasslands are dominated by diverse grass species like *Phalaris arundinacea, Agrostis stolonifera, Alopecurus pratensis*, and *Agropyron repens*, varying by land use and duration of floods. In forest and bush patches grow *Salix purpurea, Salix triandra, Salix alba*, and *Populus* spp.

2.2. Datasets

This study aims to develop a spatiotemporal analysis on the river channel development and the ecological diversity based on a set of aerial imagery in 10 different time periods (Table 1). A set of black/white military-based historical aerial photographs of the study were available from the year of 1952, 1956, 1975 and 1988 given by the Hungarian Military History Museum. The archive aerial imagery was scanned in 600 dpi resolution and then orthorectified using the ERDAS Imagine software. However, orthophotographs of 2011 were used as a reference (datum: Hungarian HD72/EOV), but we included

Topographical maps of 1980 as well since there were few field objects on the aerial imagery between 1952 and 1988, that w not possible to identify on the orthophotographs of 2011. During the process, 15–20 ground control points (GCP) were used for each image to provide accurate georeferencing. The RMSE (Root Mean Square Error) values varied between 3.4 to 6.7 m with an average of 4.8 m. We purchased digital colour orthophotographs of 2000, 2005 and 2011. The scale range of the aerial imagery and orthophotographs were found between 1:7000 and 1:12,000. After 2011, there were no official orthophotographs available from the study area. Nowadays, UAVs offer a valuable solution to produce high-resolution aerial imagery from a few km^2 large areas [50,51]; thus they are widely used for mapping wetland areas, especially for disaster management [52,53]. UAV-based surveys were conducted in 2015 by using DJI Phantom drones in order to provide orthophotographs for 2015, 2016 and 2017. Each flight was performed at low-flow conditions while, at least, 20 GCPs were measured by a Stonex RTK-GPS system. UAV image acquisition was performed at 150 m A.G.L. with 75% frontlap and sidelap between flight paths to provide a ground resolution of 0.07 to 0.09 m. Agisoft Photoscan software was used for the photogrammetric processing and the creation of orthophotographs with an overall accuracy of around 0.05 m.

Table 1. Basic parameters of aerial imagery and orthophotographs used in this study.

Year	Number of Images	Type	Scale	Resolution (m)	RMSE (m)
1952	22	B/W Aerial photo	1:7000	0.5	2.7
1956	18	B/W Aerial photo	1.7000	0.5	3.9
1975	15	B/W Aerial photo	1:12,000	0.5	2.2
1988	17	B/W Aerial photo	1:12,000	0.5	2.8
2000	22	Ortophoto	1:10,000	0.5	-
2005	22	Ortophoto	1:10,000	0.5	-
2011	22	Ortophoto	1:10,000	0.4	-
2015	1	UAV-Orthophoto	1:7498	0.09	0.05
2016	1	UAV-Orthophoto	1:8272	0.07	0.05
2017	1	UAV-Orthophoto	1:7669	0.07	0.05

2.3. Indices of River Channel Development

In order to quantify the extent of degradation by the lateral bank erosion, overlays of pairs (Figure 2) related to consecutive time periods of the river channel polygons were composed [3]. In this research, we consider that understanding the area of accretion also plays a key role in the case of ecosystem diversity analysis. These two variables were derived at the same time. The crosshatched area represents that part of the floodplain where the two-channel planforms overlap and it appears that the channel did not change position, without any pronounced erosion or accretion.

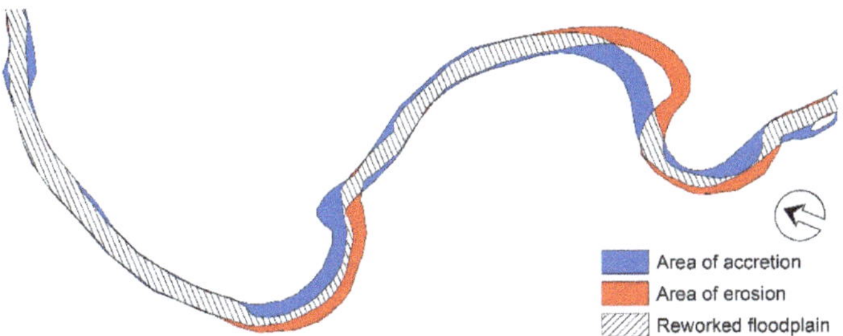

Figure 2. Methodology for calculating areas of erosion and accretion.

The value of sinuosity index (SI) was calculated as the ratio of channel length to valley length [54]; consequently, not just the areal but morphological trend of river channel development could be assessed too. We determined the channel length as the length of channel centerline while valley length was measured as the distance between the point where Sajó River crosses the Hungarian border and the point of the Tisza River confluence. We calculated total SI values for each investigated time-period. We determined the mean channel migration rate based on polygons drawn by the overlapping centerlines in each consecutive time periods. The mean lateral channel shift is a ratio of polygon area and the half of the polygon perimeter [55]. Regarding erosion/accretion and channel shift, we normalized the values by the number of years covering each different periods.

We identified the meander parameters e.g., chord length (straight line distance between two inflexion points; but this value is not equal with meander wavelength, that is the straight line distance of two consecutive meander apexes of the same side of the river banks), amplitude, the width-normalized radius of curvature (R/w). Channel widths were also determined in every 20 m along the centerlines of each time periods; therefore, a detailed mean channel width was given at the end.

2.4. Ornithological Survey

The avifauna of the study area via transect and point count surveys during the breeding season of 2018 was assessed. We selected two 200 m long line transects traversing the recently deposited areas as well as an observation point along an eroding section of the river to be able to detect birds along the full length of the study area. The transects were 500 m apart, while the point was 225 m and 580 m from the transects. We conducted the survey on 26 June between 10:00 a.m. and 12:40 p.m. Each transect survey lasted for 30 min and the point count for 20 min during which the observer recorded all species seen or heard in the vicinity of the river channel and the surrounding floodplain that are potentially using the habitat (i.e., not just flying over it at a high altitude).

We also quantified the size of the sand martin colonies of the eroding river banks of the study area from photo-mosaics. We applied close-range terrestrial photogrammetry to compose the mosaic view of the eroding river bank. For this purpose, a Nikon D5300 camera with a 70–300 mm tele-objective lens was used to capture the necessary photo-sequences of 218 images from tripod stands from the opposite side of the river. The datasets were processed in Agisoft Photoscan software. We printed coded targets of the software in A4 size in order to place them on the bank edges. We measured their precise coordinates by Stonex RTK-GPS system in order to provide a reliable scaling of the mosaic in Agisoft Photoscan.

2.5. Landscape Metrics

Class and landscape level landscape metrics to reveal the ecological features and processes along the changing river channel were applied. Initially, landscape patches were manually vectorized with a scale of 1:1000 in ArcGIS 10.3 on the aerial imagery in each investigated time-period. During the process, we have delineated the following land use categories: forests and bushes, grasslands, arable lands, bare point bar surfaces, settlement and the river channel itself (Figure 3). Aerial photographs of 1952, 1956, 1975 and 1988 are black and white ones and sometimes the quality is far from good, however, these photographs ensure a larger time range to monitor the changes. Land cover classes had to be consistent; therefore, we merged the bushes and trees of forests. We decided to apply this approach considering the four stage successional model of Corenblit [56]. In our case bare surfaces represents the vegetation recruitment phase, grasslands are in phase of vegetation establishment and succession continues with development of bushes and forests, which consists the same spectrum of species but in different age and development phase. The river channel itself is the main factor for diaspore dispersion. Arable lands and settlements are totally controlled by the society, until the migration of the river channel does not alter this situation and shift them to a previous phase of succession.

Figure 3. Aerial overview of the study area with the different land use categories (1—forests and bushes; 2—grasslands; 3—arable lands; 4—bar surfaces; 5—river channel). (**A**) Oblique aerial photo by L.B., 11 June 2017; (**B**) Black & White archive aerial photo from 1975; (**C**) UAV-based orthophoto from 2017.

In order to quantify and evaluate the direct changes in land cover and the transformation of land patches related to vegetation succession, channel migration or human agricultural management, we calculated a confusion matrix based on the vectorized land cover files in Idrisi software. In these tables, the diagonal values related to a pair of a given land cover category represents the proportion of area where no changes were found. The columns represents the initial land cover category while the rows will define the land cover categories that the given column category were transformed into. The values represents their proportion from the total area.

We concentrated on the metrics which reflect the diversity of the floodplain and its environments:

- Patch Density (PD): we calculated the index on landscape level as the number of patches per unit area (Equation (1)).

$$PD = \frac{N}{A}(10,000)(100) \tag{1}$$

where N is the number of patches in the landscape, and A is the total area (m^2), and the outcome is expressed in number· per 100 ha^{-1} [57].

- Interspersion and Juxtaposition Index (IJI): we calculated the index on landscape level as the function of observed interspersion and the maximum possible interspersion for the given number of patch types (Equation (2)).

$$\text{IJI} = \frac{-\sum_{i=1}^{m}\sum_{k=i+1}^{m}[(E_{ik}) \times ln(E_{ik})]}{ln\left(\frac{m(m-1)}{2}\right)} \tag{2}$$

where E_{ik}: the total edge between patch types i and k; m: number of patch types. IJI is expressed in per cent (0–100); low values indicate unevenly distributed or isolated patches in the area, the largest value is acquired when all patch types have a common edge with all possible other patch types [57,58].

- Shannon's Diversity Index (SHDI): We calculated the index on landscape level as the sum of the proportional abundance of each patch type multiplied by that proportions (Equation (3)).

$$\text{SHDI} = -\sum_{i=1}^{m}(P_i.lnP_i) \tag{3}$$

where P_i is the proportion of the landscape occupied by the patch type i. When the landscape is occupied by only one patch SHDI = 0 and increases with the number of patch types and their proportional area without upper limit [57,59].

- Class Area of the forests (CA_F): we considered the forests' area as the indicator of landscape change (in the initial phase, in 1956, there were only a few small patches and later, with the river bed development, the area relevantly increased). The index was calculated as the proportion of the forest land cover type and the total area and was expressed in per cent.

2.6. Statistical Analysis

After the calculation of descriptive statistics, the connection between the indices of river channel development (Sinuosity—SI, Area of erosion—Er, Area of Accretion—Ar) and the landscape metrics was also investigated. Regression analysis and Principal Component Analysis (PCA) were applied. While the regression analysis pointed on the changes directly with the given variables and provided valuable information on the temporal features and breakpoints of the surface development, PCA helped to identify the years when both the landscape and the river bed had deterministic relations.

Standardized PCA was conducted using the correlation matrix with Varimax rotation to gain orthogonal axes, i.e., non-correlating principal components (PCs) [60,61]. PD, CA_F, IJI, and SHDI landscape metrics and the Er, Acc and SI indices of riverbed development were involved. Thus, it became possible to identify both the cross-connections among the variables and to visualize the dates of different stages of surface development on a biplot diagram. Model fit was controlled with the Root Mean Square Residual (RMSR), which is determined using the residuals of the original correlation matrix and the estimation of the PCA [62]. RMSR values of <0.05 indicates very good fit [63].

We applied the Jonckheere-Terpra test to reveal whether there was a trend in river channel change over the studied period. Statistical analysis was performed using the software Past 3.19 [64] and R 3.5 [65] by applying the psych [66] and GPArotation [67] packages; furthermore, the lattice [68], the clinfun [69] and ggplot2 [70] packages were applied for the data visualization.

3. Results

3.1. Changes in Land Cover and Channel Morphology

Detailed land cover changes based on the vectorization of the available aerial imagery are shown in Figure 4. After a visual interpretation of the figure, it is clearly visible that significant changes occurred over the investigated periods (65 years). The channel developed at a remarkable rate developing three main meandering bends with high sinuosity.

According to the results of the normalized erosion and accretion rates, the first period between 1952 and 1956 showed the second highest bank erosion (0.85 ha·year^{-1}) rate (Figure 5A). Bank erosion activity radically decreased with three times by 1975. It was followed by a gentle increase between 1975–1980 but after that, it decreased again and reached the lowest bank erosion rate (0.16 ha·year^{-1}) in 2005. A notable increase started in 2010 with an outstanding maximum value of 1.12 ha·year^{-1} in the period of 2015–2016. Except for the first period (1952–1956), accretion rates followed closely the erosion rates; moreover, from the period of 2000–2005, except the above-mentioned extremely erosive year between 2015 and 2016, they exceeded their contribution over the eroded areas in the channel development. Lateral migration rates (Figure 5B) primarily followed the trend of erosion rates as it started from a relatively high rate (5.3 m) of channel shift and decreased by 1975. Its minimum

(0.8 m) was also found in 2005 then similarly started to increase, however, its peak of 2016 was not as outstanding, due to its erosion rates.

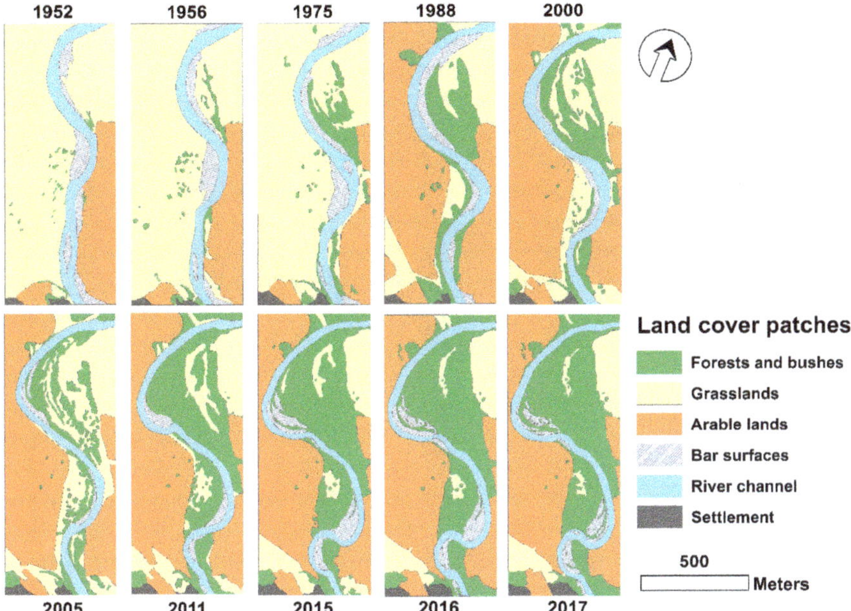

Figure 4. Land cover changes of the study area.

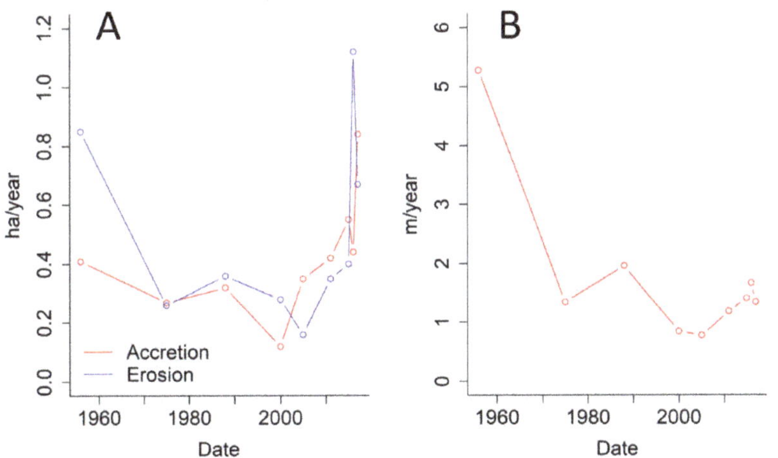

Figure 5. (**A**) Time series of normalized accretion and erosion rates in the studied period; (**B**) Time series of mean lateral channel shift rates in the studied period.

The parameters of meander evolution are summarized in Figure 6. The chord length (Figure 6A) increased in Bend 1 from 460 m to 634 m until 1988 but it was followed by a slight decrease. However, Bend 2 showed the opposite trend as it decreased to 299 m by 1988, but then the chord length increased intensely with almost 200 m by 2017. Bend 3 decreased monotonously and the total change was almost 350 m from 1952 to 2017. Amplitude (Figure 6B) increased in all the bends but the expansion of Bend 1 was almost two times higher (+127.49 m) than the others. The change in width-normalized radius of curvature (Figure 6C) showed a similar increase as it was in chord length between 1952 and 1988 but it was also found at Bend 3 as well. It was followed by an intensive decrease, especially in Bend 3.

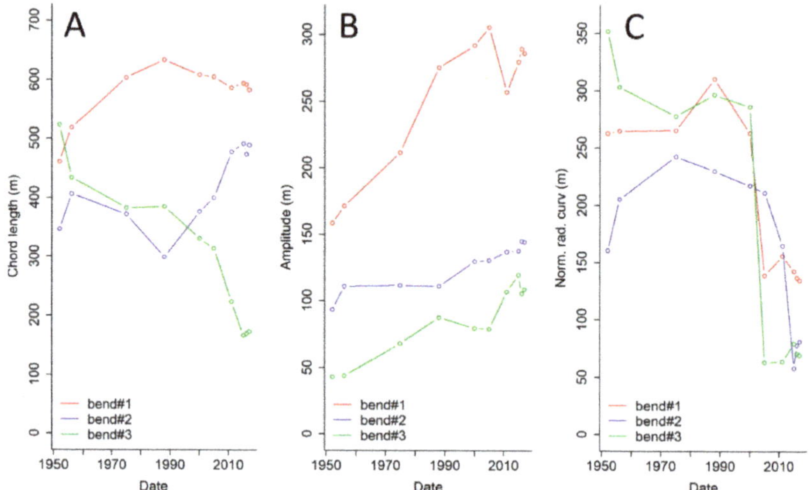

Figure 6. Parameters of meander evolution and channel shift (**A**) chord length (m); (**B**) bend amplitude (m); (**C**) width-normalized radius of curvature (m).

We justified a significant decreasing trend in case of river channel width (Figure 7) as a function of time (J-T statistic = 80298; $p = 0.0004$) and there was a negative correlation between the mean channel width and sinuosity ($r = -0.93$, $p < 0.001$).

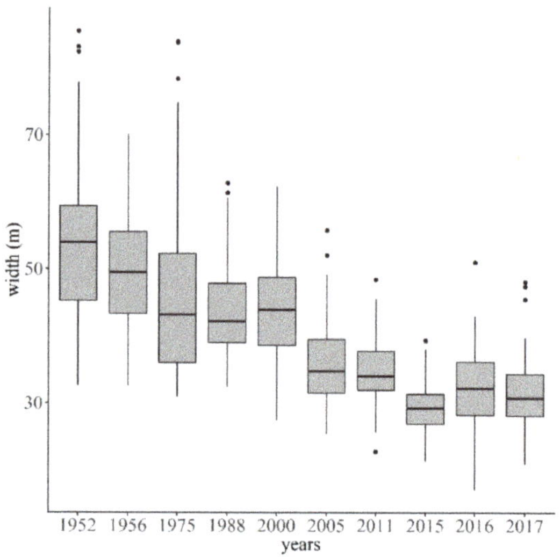

Figure 7. Changes in the mean channel width.

Land cover changes can be divided into two groups: (1) settlement, river channel and point bars where the changes were minimal (the area was more or less constant but the patches had spatial changes); (2) arable land, grassland and forests where the changes were relevant (these patch types turned into another, at least, partly). The area of grasslands decreased transforming into arable land and forest (Figure 8). The extent of the forests and bushes increased by 25% since the share of grasslands showed a decrease in similar rate and bare surfaces were not changed considerably. The extent of human-controlled arable lands was increased at the beginning but slightly decreased in the last decades. Extents of the river channel and settlement were not changed considerably.

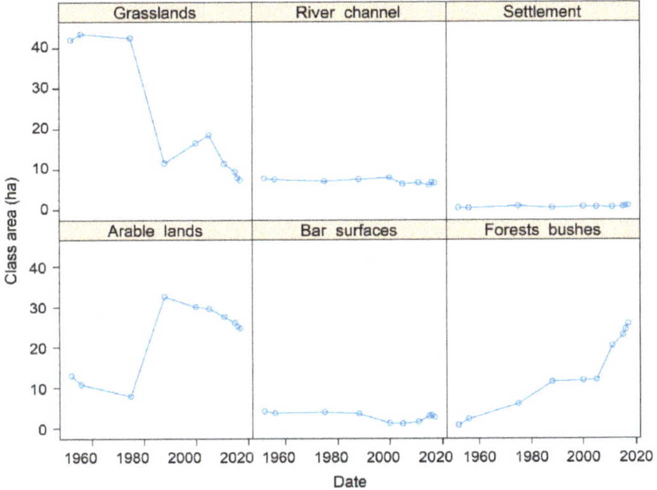

Figure 8. Changes in land cover over the studied period by patch types.

The detailed contingency tables of land cover changes between each consecutive time periods were summarized and available in Supplementary File S1. In Table 2, we presented the changes of selected pairs of land use category transformations that showed higher conversion proportions. The first four rows represent the successional phases that follow the channel migration: The second highest transformation rates were connected to the transformation from former river channel to bar surfaces. It was followed by colonization of the bar surfaces by grasslands where the transformation rates were found also lower and totally stopped by 2005. However, generally, direct transformation from bar surfaces to forests showed some higher values but it could have been expressed by the longer time periods covered. The highest transformation rates were found in the transformation from grasslands to forests. The last four rows represents transformation categories that were controlled by the channel migration and bank retreat except the changes from grasslands to arable lands. This type also resulted in an outstanding proportion (0.326) between 1975 and 1988 and it radically decreased afterwards.

Table 2. Conversion matrix (changes in percent) of LC categories in the investigated periods.

Conversion Type		1952–1956	1956–1975	1975–1988	1988–2000	2000–2005	2005–2011	2011–2015	2015–2016	2016–2017
Initial	**Final**									
RC	BS	0.016	0.019	0.028	0.007	0.013	0.014	0.018	0.005	0.004
BS	G	0.008	0.016	0.005	0.008	0.002	0	0	0	0
BS	F	0.002	0.005	0.021	0.026	0.012	0.013	0.006	0.005	0.009
G	F	0.010	0.030	0.049	0.008	0.023	0.088	0.030	0.014	0.007
G	RC	0.008	0.025	0.042	0	0.001	0.005	0.001	0	0
G	AL	0.001	0.008	0.326	0.002	0.006	0.010	0.007	0.001	0.001
F	RC	0.001	0.002	0.004	0.007	0.003	0.004	0.004	0.006	0.002
AL	RC	0.011	0.011	0	0.017	0.006	0.018	0.015	0.004	0.004

AL—Arable lands, BS—Bar surfaces, F—Forests, G—Grasslands, RC—River channel.

A linear relationship between the increase of forest areas and the SI ($R^2 = 0.94$, $p < 0.001$; Figure 9A) was revealed. At the first date of the study period (1956), only a few patches of the forest were observed, which reached a 25% proportion till 2017; besides, SI indicated a straight river channel section in 1956 and it became a complex form up to 2017. Thus, the forest areas increased directly as the river channel transformed and appropriate areas developed in the floodplain.

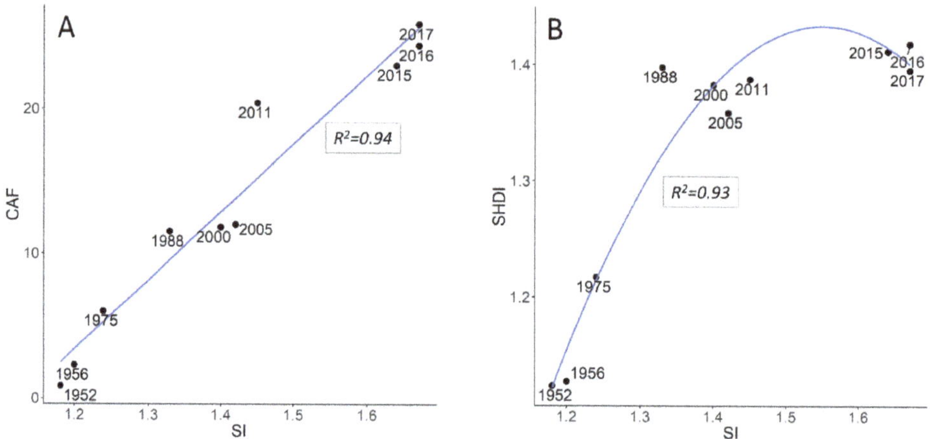

Figure 9. Sinuosity as independent factor of land cover changes and landscape diversity in the study area. (**A**) The relationship between the SI (Sinuosity Index) and CAF (Class area of Forests); (**B**) Relationship between the SI (Sinuosity Index) and SHDI (Shannon's Diversity Index) metrics.

There was a significant connection between the SI and SHDI variables ($R^2 = 0.93$, $p < 0.001$). However, unlike in the case of forest areas, the connection was not linear (Figure 7B); a second-order polynomial regression indicated that there was a change in the trend after 2011: next years (2015–2017) were outside of the linear trend and the curve showed saturation; moreover, there was a small decrease in the diversity in 2017.

PCA explained the 94% of the total variance. Three principal components (PC) were justified by the RMRS (0.03). PC1 correlated with the Er (Area of Erosion), Acc (Area of Accumulation) and PD (Patch Density) explaining 40% of the total variance. PC2 correlated with the SI, SHDI, and CAF and accounted for 37% of the total variance, while PC3 was formed by solely the IJI having 18% in the explanation of the total variance. Acc and Er correlated with the PD and SI with the SHDI and CAF.

The ordination diagram showed that years formed two distinct groups differentiated by the Er, Acc and PD indices (Figure 10). The first group was situated in the negative part of the diagram, and the second, with a larger range, in the positive part. Considering the vertical axis (PC2), the distribution followed a monotonous increasing trend between 1956 and 1988 but it became sparse after 2000 with 2005 having the lowest value and after another increase, it started to decrease from 2015.

3.2. Avifauna

Since the biodiversity can be affected by the different land cover and morphological river changes, we surveyed the avifauna as a possible bioindicator. Result of the snapshot faunistic survey is shown in Table 3. A total of 26 species were detected breeding or using the habitat during the breeding season, of which 23 are protected under the Hungarian law including the strictly protected European bee-eater (*Merops apiaster*) [71].

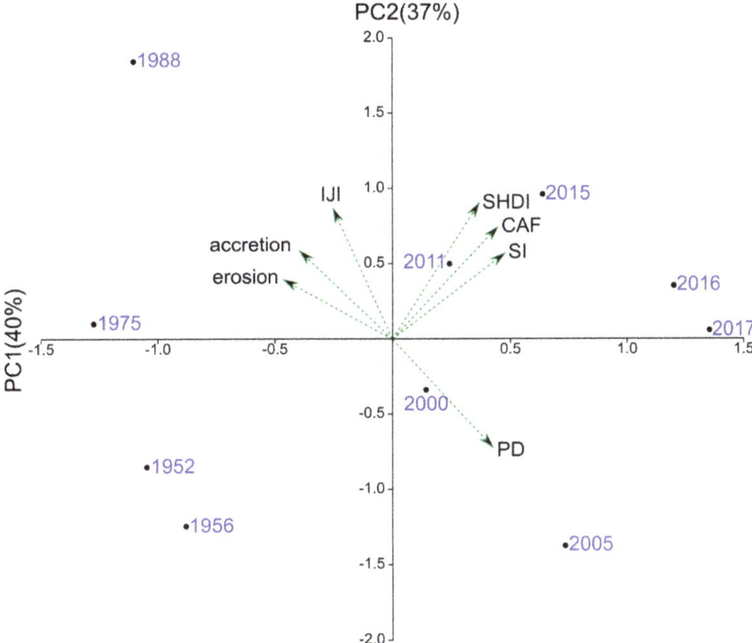

Figure 10. Biplot of the PCA model conducted with the landscape metrics and river channel development indices (dashed arrows: involved variables).

Table 3. List of bird species seen and/or heard during the censuses listed in taxonomic order.

	Common Name	Scientific Name
1	Grey heron	*Ardea cinerea*
2	Common buzzard	*Buteo buteo*
3	Common kestrel	*Falco tinnunculus*
4	Eurasian hobby	*Falco subbuteo*
5	Common pheasant	*Phasianus colchicus*
6	Common sandpiper	*Tringa hypoleucos*
7	Woodpigeon	*Columba palumbus*
8	Turtle dove	*Streptopelia turtur*
9	Common cuckoo	*Cuculus canorus*
10	Common kingfisher	*Alcedo atthis*
11	European bee-eater	*Merops apiaster*
12	Green woodpecker	*Picus viridis*
13	Great spotted woodpecker	*Dendrocopus major*
14	Sand martin	*Riparia riparia*
15	Common blackbird	*Turdus merula*
16	River warbler	*Locustella fluviatilis*
17	Eurasian blackcap	*Sylvia atricapilla*
18	Chiffchaff	*Phylloscopus collybita*
19	Great tit	*Parus major*
20	Common starling	*Sturnus vulgaris*
21	Golden oriole	*Oriolus oriolus*
22	Eurasian jay	*Garrulus glandarius*
23	Hooded Crow	*Corvus cornix*
24	Common chaffinch	*Fringilla coelebs*
25	European greenfinch	*Carduelis chloris*
26	European goldfinch	*Carduelis carduelis*

The nest cavities of both the bee-eater and the steeply declining Sand martin (*Riparia riparia*) was found along the eroding section of the Sajó River bank. Approximately, 80–100 pairs of sand martins were feeding young at the time of the survey at the focal section of the river. However, based on the number of visible nest cavities (Figure 11), the two colonies could have consisted of a total of 446 breeding pairs.

Figure 11. Nest cavities along the eroding river banks of Sajó River at the study area. (**A**) The overview of the cavities; (**B**) Examples of cavities along the recently eroded slump blocks; (**C**) Red circles indicates the cavities identified on one part of the mosaic.

4. Discussion

Fluvial morphological changes are able to generate intensive modifications, which can affect hydrological processes and, even, biodiversity. One clear example is the Sajó River, which was assessed in this research. Sajó River itself is regulated only in particular reaches, but its morphological evolution is affected by the Tisza River. Sajó River as a tributary joins Tisza River and the study area is located only about 20 km upstream from the confluence, which is strongly affects the sediment transport and channel development of the studied Sajó River sub-reach. The Tisza was regulated during the second half of the 19th century, and as a result of these regulations, its riverbed was incised for its straightened channel and increased energy [42,43]. This process could be also responsible for the morphological changes of Sajó River. The characteristics of Tisza channel development basically changed after the construction dams on Tisza River, both on upstream (1954, Tiszalök) and on downstream (1978, Kisköre) sections around the firth of Sajó. The establishment of these dams could cause a reduction of sediment supply [54] which generally leads to further channel incision. However, we determined the narrowing (Figure 7), and therefore, similarly to Tisza River, a possible incision, of the Sajó River but its channel started to develop in a different way, namely increased meandering started instead of a former incision, which has diverse effects on land cover, conservational value and productivity on surrounding landscape [72].

Morphological changes allow registering high rates of bank erosion, which in similar investigations in other European rivers, was also calculated [3,73,74]; however, the extent of arable lands lost by lateral erosion is not an outstanding value compared to other rivers with similar geomorphological patterns [75]. In this way, it would be a great opportunity to include in future investigations the correlation between these morphological alterations, the changes in the biodiversity and the bank erosion rates.

It is notable that the accretion of the river bank mainly followed the rate of eroding outer banks; therefore, new bare point bar surfaces had been developed in a short time period. Although these surfaces are not valuable ecologically, at the beginning phase, they represent potential future habitats [74]; our study pointed out that in case the vegetation can rapidly occupy the area, then the process can increase the extent of habitat in the ecological corridors. Moreover, the studied sub-reach is situated between two Natura2000 areas, thus, channel migration can enhance the habitat diversity and species connectivity between these sub-groups of the ecological network, that strengthen the role of an ecological corridor, as a possible compensation of the land degradation processes.

The majority of previous studies were focusing on the extensive negative effects of river bank erosion i.e., spreading pollutants from upstream reservoirs or remobilizing heavy metal contaminants [76–78] but only a few discuss their ecological consequences. In some cases, under different climate conditions, the lateral shifts of river channels rapidly decrease the biodiversity of the neighbouring flora and fauna [79] but in this case, it affects only agricultural areas. In our study, we demonstrated an opposite dynamic: How the process can be valuable. Regarding the changes in landscape metrics, as SI values increased, the forest areas (CA-F) also increased. SI reflected how the planform of the river channel was developed, and the positive connection with the forest areas indicate that instead of the arable lands, a natural afforestation process was initiated. Small grassland patches can be merged into forest cover without appropriate management [80] but in this case, the increase of SI provided the criteria of the forest cover extension. The newly developed point bars were occupied by plants (firstly with herbaceous plants) and later became forested following a successional process. Diversity did not follow the linear trend because the area of the forests reached a threshold when the further increase did not increase the diversity of the mosaic of land cover patches (Figure 9). This is the explanation of the two kinds of the relationship of the SI with the landscape metrics. The analysis on the conversion matrix between land cover categories showed higher values on the changes related to vegetation succession phases while the lower values were found in connection with the channel migration. However, we found outstanding transformation from grasslands to arable lands between 1975 and 1980 but this can be clearly explained by human intervention on the land management.

Multivariate analysis revealed that 1956–1988 period had rather similar features considering the erosion-accretion processes and the PD, and there was another group of years between 2000 and 2017 that can be distinguished according to the remaining variables. However, considering the results of PCA, SI, SHDI, and CA_F variables differentiated the points a sparse distribution: As the erosion and accretion processes increased or decreased, or were larger or smaller compared to each other, the forest areas and the diversity also reacted differently. This process became the major regulator of the ecological process from 1988 to 2000 when the river channel development reached SI = 1.4 and in 2015 when it reached SI = 1.6, the pace of the lateral movement slowed down. At the beginning of meander development, the chord length remains stable and an extension, with increasing amplitude, occurs [11]. We observed similar changes along the studied Sajó River reach since the chord lengths showed only minor changes until 1975 from where both increase (bend 2) and decrease (bend 1, 3) took place. According to [13] the decreasing normalized curvature (R/w) accelerates the channel migration as it was also found (Figures 5b and 6c) in the studied Sajó River reach. The proportion of the forests increased from 1 to 25%, and, what is important, these areas were constant (if a patch type transformed into forest, most of the area remained forest in the consecutive years, too), while the grasslands, arable lands, and especially the point bars changed in spatial terms (changes can occur back and forth, or the formation and diminishing is a common process).

The development of successional riparian and fluvial forests on recently deposited point bars significantly increases biodiversity locally [81] but more importantly, it facilitates the movement of organisms across an otherwise cultivated and homogenous landscape, thus, allows the maintenance of gene flow between meta-populations. Indeed, the classification of the study area is "ecological corridor" within the National Ecological Network, which connects ecologically important core areas along the river. The avifauna of the study area is similar to communities detected along with other reaches of the river over two decades ago [82] suggesting that the newly formed habitats are quickly colonized by protected species from nearby areas.

The eroding banks provide important nesting sites for colonies of protected and regionally declining migratory bird species such as the sand martin and the European bee-eater [83] further increasing the ecological importance of the new habitats created by this dynamic river. In fact, an effective way to improve the availability of nesting sites and facilitate the recolonization of an area by sand martins is the removal of bank erosion control projects from heavily regulated river channels [84]. This may be particularly important for maintaining viable populations of sand martins in agricultural landscapes as these birds usually do not reuse their nests from previous years to avoid the costs of heavy infestation with parasites [85,86], rather select nesting sites along river banks and sand quarries where periodically renewing vertical surfaces are available at the beginning of the nesting season. Eroding river banks at our study area provide just that, the continuous renewal of this critical resource, the nesting site, which often limits the distribution of both sand martins and bee-eaters [71,85].

The increase in the extent of the most stable successive landscape element (forests) and a decrease of elements in middle-phase of succession (grasslands) suggest that the land-mosaic alteration during the investigated time resulted in higher diversity and stability. This process is important regarding laterally active channels; therefore, identification of the relationship between morphological changes in river channel and landscape evolution is vital [2,56]. Changes of arable lands could not be used as an indication for river development since it is controlled by human actions related to agricultural management. As a main controlling factor for successive landscape elements, the altered dynamic of meandering could be evaluated. Even if no relevant human interventions were present within the study area, the changes in meandering-dynamic can be highly influenced by the anthropogenic changes in the flood-dynamic of the main stem Tisza River. Decreasing agricultural productivity due to bank failures and channel shifting seems to be balanced by increasing habitat diversity by recent point bar deposits, succession areas, and bushes. Considering the increasing share of cultivated areas during last decades within a floodplain, which is part of a national ecological network, positive effects of the meandering of the river exceeds the negative effects of land loss and lateral erosion.

Another important topic to be assessed in the future would be the connectivity processes [87]. It is clear that the studied river and surrounding landscapes are connected by different processes such as nutrient transport or sediment mobilization, both of these are also conditioned by the vegetation colonization [88–90]. The application of modelling techniques and connectivity indexes allow us detecting how meandering is influenced by other dynamic fluxes such as agricultural fields [91] or urban areas [92] and at which level. Therefore, undoubtedly, understanding connectivity processes in the Sajó river would help land plan managers and stakeholders to design correct and sustainable soil erosion control measures and water conservation practices, as other authors also confirmed in other degraded areas [93,94].

5. Conclusions

River channel development is a natural process, which usually considered to have a negative effect due to the damages caused to agriculture or infrastructure. Our hypothesis was that, besides the detrimental phase, there is also a positive effect because of the newly developed habitats on the opposite side of the river. We revealed that 65 years were enough to gain a new habitat system along the river as the linear channel formed into a meandering and more natural state. At the beginning phase, there were only a few patches of forests and the matrix had been dominated by the surrounding arable lands, while nowadays the forests have a major role in the landscape mosaic. There was a linear relationship between the sinuosity and the class area of the forests; i.e., the more developed the meanders were, the more forest patches appeared in the area. However, Shannon's diversity did not follow a linear trend with sinuosity and instead of reaching its maximum it showed a polynomial trend; i.e., the areal increase of forests did not increase the landscape diversity as the patches became dominant, compact and connected. Consequently, this was an advantageous process from ecological aspects. Although the plant species were not of significant conservation value (mostly pioneers and weeds) but provided habitats for several protected bird species. Besides, the eroding side of the riverbed also serves as a nesting place for birds, too. Accordingly, we emphasize the positive effects of the erosion and accretion processes, as nature conservation benefits from the new geomorphological forms of river channel development.

Supplementary Materials: The Supplementary Materials are available online at http://www.mdpi.com/2073-4441/10/11/1613/s1.

Author Contributions: L.B. designed and carried out this research; S.S. and Á.K. supervised and instructed this research; L.B., S.S., T.J.N., and Z.N. prepared and analyzed the data. L.B. lead the writing with contributions from S.S., T.J.N., Z.N. and J.R.-C.; Á.K. did the proofreading. All the authors have approved the manuscript.

Funding: The research was supported by the National Research, Development and Innovation Office (NKFIH; 108755). L.B. was supported by the ÚNKP-17-3 New National Excellence Program of the Ministry of Human Capacities. The project was supported by the European Union, co-financed by the European Regional Development Fund (EFOP-3.6.1-16-2016-00022).

Conflicts of Interest: The authors declare no conflict of interest.

References

1. Hooke, J.M. Temporal variations in fluvial processes on an active meandering river over a 20-year period. *Geomorphology* **2008**, *100*, 3–13. [CrossRef]
2. Michalková, M.; Piégay, H.; Kondolf, G.M.; Greco, S.E. Lateral erosion of the Sacramento River, California (1942–1999), and responses of channel and floodplain lake to human influences. *Earth Surf. Process. Landf.* **2010**, *36*, 257–272. [CrossRef]
3. Rusnák, M.; Lehotský, M. Time-focused investigation of river channel morphological changes due to extreme floods. *Z. für Geomorphol.* **2014**, *58*, 251–266. [CrossRef]
4. Konsoer, K.M.; Rhoads, B.L.; Best, J.L.; Langendoen, E.J.; Abad, J.D.; Parsons, D.R.; Garcia, M.H. Three-dimensional flow structure and bed morphology in large elongate meander loops with different outer bank roughness characteristics. *Water Resour. Res.* **2016**, *52*, 9621–9641. [CrossRef]
5. Langendoen, E.J.; Andrew, S. Modeling the Evolution of Incised Streams. II: Streambank Erosion. *J. Hydraul. Eng.* **2008**, *134*, 905–915. [CrossRef]
6. Dragićević, S.; Pripužić, M.; Živković, N.; Novković, I.; Kostadinov, S.; Langović, M.; Milojković, B.; Čvorović, Z. Spatial and temporal variability of bank erosion during the period 1930–2016: Case study—Kolubara River Basin (Serbia). *Water* **2017**, *9*, 748. [CrossRef]
7. Nakano, D.; Nakamura, F. The significance of meandering channel morphology on the diversity and abundance of macroinvertebrates in a lowland river in Japan. *Aquat. Conserv. Mar. Freshw. Ecosyst.* **2007**, *18*, 780–798. [CrossRef]
8. Kiss, T.; Balogh, M. Characteristics of Point-Bar Development under the Influence of a Dam: Case Study on the Dráva River at Sigetec, Croatia. *J. Environ. Geogr.* **2015**, *8*, 23–30. [CrossRef]
9. Garcia, X.F.; Schnauder, I.; Pusch, M.T. Complex hydromorphology of meanders can support benthic invertebrate diversity in rivers. *Hydrobiologia* **2012**, *685*, 49–68. [CrossRef]
10. Szabó, Z.; Tóth, C.A.; Tomor, T.; Szabó, S. Airborne LiDAR point cloud in mapping of fluvial forms: A case study of a Hungarian floodplain. *GISci. Remote Sens.* **2017**, *54*, 862–880. [CrossRef]
11. Hickin, E.J. The Development of Meanders in Natural River Channels. *Am. J. Sci.* **1974**, *274*, 414–442. [CrossRef]
12. Ward, J.V.; Tockner, K.; Arscott, D.B.; Claret, C. Riverine landscape diversity. *Freshw. Biol.* **2002**, *47*, 517–539. [CrossRef]

13. Hickin, E.J. Hydraulic factors controlling channel migration. In *Research in Fluvial Geomorphology, Proceedings of the Fifth Guelph Symposium on Geomorphology*; Davidson-Arnott, R., Nickling, W., Eds.; Geo Abstracts Ltd.: Norwich, UK, 1978; pp. 59–66.

14. Gurnell, A.M.; Grabowski, R.C. Vegetation—Hydrogeomorphology Interactions in a Low-Energy, Human-Impacted River. *River Res. Appl.* **2015**, *32*, 202–215. [CrossRef]

15. Gurnell, A.M.; Bertoldi, W.; Corenblit, D. Changing river channels: The roles of hydrological processes, plants and pioneer fluvial landforms in humid temperate, mixed load, gravel bed rivers. *Earth-Sci. Rev.* **2012**, *111*, 129–141. [CrossRef]

16. Shin, N.; Nakamura, F. Effects of fluvial geomorphology on riparian tree species in Rekifune River, northern Japan. *Plant Ecol.* **2005**, *178*, 15–28. [CrossRef]

17. Sandercock, P.J.; Hooke, J.M.; Mant, J.M. Vegetation in dryland river channels and its interaction with fluvial processes. *Prog. Phys. Geogr. Earth Environ.* **2007**, *31*, 107–129. [CrossRef]

18. Kleinhans, M.G.; de Vries, B.; Braat, L.; van Oorschot, M. Living landscapes: Muddy and vegetated floodplain effects on fluvial pattern in an incised river. *Earth Surf. Process. Landf.* **2018**. [CrossRef]

19. Zeng, Q.; Shi, L.; Wen, L.; Chen, J.; Duo, H.; Lei, G. Gravel Bars Can Be Critical for Biodiversity Conservation: A Case Study on Scaly-Sided Merganser in South China. *PLoS ONE* **2015**, *10*, e0127387. [CrossRef] [PubMed]

20. Robertson, K.M. Distributions of tree species along point bars of 10 rivers in the south-eastern US Coastal Plain. *J. Biogeogr.* **2005**, *33*, 121–132. [CrossRef]

21. Wintenberger, C.L.; Rodrigues, S.; Bréhéret, J.G.; Villar, M. Fluvial islands: First stage of development from nonmigrating (forced) bars and woody-vegetation interactions. *Geomorphology* **2015**, *246*, 305–320. [CrossRef]

22. Cotton, J.A.; Wharton, G.; Bass, J.A.B.; Heppell, C.M.; Wotton, R.S. The effects of seasonal changes to in-stream vegetation cover on patterns of flow and accumulation of sediment. *Geomorphology* **2006**, *77*, 320–334. [CrossRef]

23. Gurnell, A.M.; Corenblit, D.; García de Jalón, D.; González del Tánago, M.; Grabowski, R.C.; O'Hare, M.T.; Szewczyk, M. A Conceptual Model of Vegetation–hydrogeomorphology Interactions within River Corridors. *River Res. Appl.* **2015**, *32*, 142–163. [CrossRef]

24. Hackney, C.; Best, J.; Leyland, J.; Darby, S.E.; Parsons, D.; Aalto, R.; Nicholas, A. Modulation of outer bank erosion by slump blocks: Disentangling the protective and destructive role of failed material on the three-dimensional flow structure. *Geophys. Res. Lett.* **2015**, *42*. [CrossRef]

25. Konsoer, K.; Rhoads, B.; Best, J.; Langendoen, E.; Ursic, M.; Abad, J.; Garcia, M. Length scales and statistical characteristics of outer bank roughness for large elongate meander bends: The influence of bank material properties, floodplain vegetation and flow inundation. *Earth Surf. Process. Landf.* **2017**, *42*, 2024–2037. [CrossRef]

26. Zawiejska, J.; Wyżga, B. Twentieth-century channel change on the Dunajec River, southern Poland: Patterns, causes and controls. *Geomorphology* **2010**, *117*, 234–246. [CrossRef]

27. Hohensinner, S.; Jungwirth, M.; Muhar, S.; Schmutz, S. Spatio-temporal habitat dynamics in a changing Danube River landscape 1812–2006. *River Res. Appl.* **2011**, *27*, 939–955. [CrossRef]

28. Kiss, T.; Blanka, V. River channel response to climate- and human-induced hydrological changes: Case study on the meandering Hernád River, Hungary. *Geomorphology* **2012**, *175–176*, 115–125. [CrossRef]

29. Nakamura, F.; Yamada, H. Effects of pasture development on the ecological functions of riparian forests in Hokkaido in northern Japan. *Ecol. Eng.* **2005**, *24*, 539–550. [CrossRef]

30. Campana, D.; Marchese, E.; Theule, J.I.; Comiti, F. Channel degradation and restoration of an Alpine river and related morphological changes. *Geomorphology* **2014**, *221*, 230–241. [CrossRef]

31. Habersack, H.; Hein, T.; Stanica, A.; Liska, I.; Mair, R.; Jäger, E.; Hauer, C.; Bradley, C. Challenges of river basin management: Current status of, and prospects for, the River Danube from a river engineering perspective. *Sci. Total Environ.* **2016**, *543*, 828–845. [CrossRef] [PubMed]

32. Hajdukiewicz, H.; Wyżga, B. Aerial photo-based analysis of the hydromorphological changes of a mountain river over the last six decades: The Czarny Dunajec, Polish Carpathians. *Sci. Total Environ.* **2019**, *648*, 1598–1613. [CrossRef] [PubMed]

33. Negishi, J.N.; Inoue, M.; Nunokawa, M. Effects of channelisation on stream habitat in relation to a spate and flow refugia for macroinvertebrates in northern Japan. *Freshw. Biol.* **2002**, *47*, 1515–1529. [CrossRef]

34. Kondolf, G.M. River restoration and meanders. *Ecol. Soc.* **2006**, *11*, 42. [CrossRef]

35. Palmer, M.A.; Hondula, K.L.; Koch, B.J. Ecological Restoration of Streams and Rivers: Shifting Strategies and Shifting Goals. *Annu. Rev. Ecol. Evol. Syst.* **2014**, *45*, 247–269. [CrossRef]

36. Clark, M.J.; Montemarano, J.J. Short-Term Impacts of Remeandering Restoration Efforts on Fish Community Structure in a Fourth-Order Stream. *Water* **2017**, *9*, 546. [CrossRef]

37. Jähnig, S.C.; Brabec, K.; Buffagni, A.; Erba, S.; Lorenz, A.W.; Ofenböck, T.; Verdonschot, P.F.M.; Hering, D. A comparative analysis of restoration measures and their effects on hydromorphology and benthic invertebrates in 26 central and southern European rivers. *J. Appl. Ecol.* **2010**, *47*, 671–680. [CrossRef]

38. Palmer, M.A.; Menninger, H.L.; Bernhardt, E. River restoration, habitat heterogeneity and biodiversity: A failure of theory or practice? *Freshw. Biol.* **2010**, *55*, 205–222. [CrossRef]

39. Ihrig, D. *A magyar vízszabályozás története (History of the Hungarian River Regulations)*; Akadémiai Kiadó: Budapest, Hungary, 1973.

40. Dunka, S.; Fejér, L.; VágáS, I. *A verítékes honfoglalás—A Tisza szabályozás története (The New Concquest—History of The Regulation of Tisza River)*; Vízügyi Múzeum és Levéltár: Budapest, Hungary, 1996.

41. Kiss, T.; Fiala, K.; Sipos, G. Alterations of channel parameters in response to river regulation works since 1840 on the Lower Tisza River (Hungary). *Geomorphology* **2008**, *98*, 96–110. [CrossRef]

42. Amissah, G.; Kiss, T.; Fiala, K. Morphological Evolution of the Lower Tisza River (Hungary) in the 20th Century in Response to Human Interventions. *Water* **2018**, *10*, 884. [CrossRef]

43. Bertalan, L.; Rodrigo-Comino, J.; Surian, N.; Šulc Michalková, M.; Szabó, G. Complex assessment of channel changes and bank erosion hazard on the Sajó (Slaná) River, Hungary. In *Geomorfologický sborník 16, Proceedings of the Conference: State of Geomorphological Research in 2018, Vílanec, Czech Republic, 25–27 April 2018*; Máčka, Z., Ježková, J., Nováková, E., Kuda, F., Eds.; Masaryk University: Brno, Czech Republic, 2018; pp. 13–14.

44. Bogárdi, J. A Sajó hordalékszállítása és a hordalékos víz ülepítése. (Sediment transport and deposition of Sajó River). *Hidrológiai Közlöny/Hung. J. Hydrol.* **1949**, *29*, 376–379.

45. Kákóczki, B. *A Szederkényi Uradalom Történeti Földrajza*, 1st ed.; Tiszaújváros város Önkormányzata a Derkovits Gyula Művelődési Központ közreműködésével: Tiszaújváros, Hungary, 2016.

46. Bertalan, L.; Szabó, G. Lateral erosion monitoring along a southern section of Sajó (Slaná) River. In *Detailed Aerial Mapping and Flood Impact Monitoring in the V4 Region*; Křížová, A., Ed.; Univerzita Komenskeho, Bratislava: Bratislava, Slovakia, 2015; p. 4.

47. Bertalan, L.; Szabó, G.; Szabó, S. Soil degradation induced by lateral erosion of a non-regulated alluvial river (Sajó River, Hungary). In *Aktuální Environmentálni Hrozby a Jejich Impakt v Krajiné (Current Environmental Threats and Their Impact in the Landscape Brno): Sbornik Abstraktu Z Mezinárodniho Workshopu*; Zapletalová, J., Kirchner, K., Eds.; Ústav geoniky AV ČR: Poruba, Czech Republic, 2016; pp. 8–9.

48. Jongman, R.H.G.; Bouwma, I.M.; Griffioen, A.; Jones-Walters, L.; Van Doorn, A.M. The Pan European Ecological Network: PEEN. *Landsc. Ecol.* **2011**, *26*, 311–326. [CrossRef]

49. Larned, S.T.; Datry, T.; Arscott, D.B.; Tockner, K. Emerging concepts in temporary-river ecology. *Freshw. Biol.* **2010**, *55*, 717–738. [CrossRef]

50. Rusnák, M.; Sládek, J.; Kidová, A.; Lehotský, M. Template for high-resolution river landscape mapping using UAV technology. *Measurement* **2018**, *115*, 139–151. [CrossRef]

51. Szabó, G.; Bertalan, L.; Barkóczi, N.; Kovács, Z.; Burai, P.; Lénárt, C. Zooming on Aerial Survey. In *Small Flying Drones: Applications for Geographic Observation*; Casagrande, G., Sik, A., Szabó, G., Eds.; Springer International Publishing: Cham, Switzerland, 2018; pp. 91–126, ISBN 978-3-319-66577-1.

52. Restás, Á. Drone Applications for Supporting Disaster Management. *World J. Eng. Technol.* **2015**, *3*, 316–321. [CrossRef]

53. Restás, Á. Water Related Disaster Management Supported by Drone Applications. *World J. Eng. Technol.* **2018**, *6*, 116–126. [CrossRef]

54. Brierley, G.J.; Fryirs, K.A. *Geomorphology and River Management: Applications of the River Styles Framework*; Blackwell Publishing: Hoboken, NJ, USA, 2005; ISBN 1405115165.

55. Micheli, E.R.; Kirchner, J.W.; Larsen, E.W. Quantifying the effect of riparian forest versus agricultural vegetation on river meander migration rates, central Sacramento River, California, USA. *River Res. Appl.* **2004**, *20*, 537–548. [CrossRef]

56. Corenblit, D.; Tabacchi, E.; Steiger, J.; Gurnell, A.M. Reciprocal interactions and adjustments between fluvial landforms and vegetation dynamics in river corridors: A review of complementary approaches. *Earth-Sci. Rev.* **2007**, *84*, 56–86. [CrossRef]

57. McGarigal, K.; Marks, B. FRAGSTATS: Spatial Pattern Analysis Program for Quantifying Landscape Structure. *Gen. Tech. Rep. PNW-GTR-351. USDA* **1995**, *122*, 351.

58. Lopez, R.R.D.; Frohn, R.C. *Remote Sensing for Landscape Ecology: New Metric Indicators: Monitoring, Modeling, and Assessment of Ecosystems*, 2nd ed.; CRC Press: Boca Raton, FL, USA, 2017.

59. Hill, M.O. Diversity and Evenness: A Unifying Notation and Its Consequences. *Ecology* **1973**, *54*, 427–432. [CrossRef]

60. Davis, J.C. *Statistics and Data Analysis in Geology*; Wiley: Hoboken, NJ, USA, 1986; ISBN 978-0471172758.

61. Kaiser, H.F. The varimax criterion for analytic rotation in factor analysis. *Psychometrika* **1958**, *23*, 187–200. [CrossRef]

62. Joreskog, K.; Sorbom, D. *LISREL 8 User's Reference Guide*; Scientific Software International: Chicago, IL, USA, 1993.

63. Basto, M.; Pereira, J.M. An SPSS R-Menu for Ordinal Factor Analysis. *J. Stat. Softw.* **2012**, *46*, 1–29. [CrossRef]

64. Hammer, Ø.; Harper, D.A.T.A.T.; Ryan, P.D. PAST: Paleontological Statistics Software Package for Education and Data Analysis. *Palaeontol. Electron.* **2001**. [CrossRef]

65. R Core Team R: A Language and Environment for Statistical Computing. R Found. Statistical Computing Vienna Austria. 2018. Available online: http://www.R-project.org/ (accessed on 10 September 2018).

66. Revelle, W. psych: Procedures for Personality and Psychological Research. *R Package* **2016**. [CrossRef]

67. Bernaards, C.A.; Jennrich, R.I. Gradient projection algorithms and software for arbitrary rotation criteria in factor analysis. *Educ. Psychol. Meas.* **2005**, *65*, 676–696. [CrossRef]

68. Sarkar, D. *Lattice: Multivariate Data Visualization with R*; Springer: New York, NY, USA, 2008; ISBN 9780387759692r0387759697.

69. Seshan, V.E. clinfun: Clinical Trial Design and Data Analysis Functions. *R Package*, 2018. Version 1.0.15. Available online: https://CRAN.R-project.org/package=clinfun (accessed on 10 September 2018).

70. Wickham, H. *ggplot2: Elegant Graphics for Data Analysis*; Springer: New York, NY, USA, 2009; ISBN 978-0-387-98140-6.

71. Gyurácz, J.; Nagy, K.; Fuisz, T.I.; Karcza, Z.; Szép, T. European bee-eater (*Merops apiaster* Linnaeus, 1758) in Hungary: A review. *Ornis Hung.* **2013**, *21*, 1–22. [CrossRef]

72. Cserkész-Nagy, Á.; Tóth, T.; Vajk, Ö.; Sztanó, O. Erosional scours and meander development in response to river engineering: Middle Tisza region, Hungary. *Proc. Geol. Assoc.* **2010**, *121*, 238–247. [CrossRef]

73. Ondruch, J.; Máčka, Z. Response of lateral channel dynamics of a lowland meandering river to engineering-derived adjustments—An example of the Morava River (Czech Republic). *Open Geosci.* **2015**, *7*, 588–605. [CrossRef]

74. Lotsari, E.; Vaaja, M.; Flener, C.; Kaartinen, H.; Kukko, A.; Kasvi, E.; Hyyppä, H.; Hyyppä, J.; Alho, P. Annual bank and point bar morphodynamics of a meandering river determined by high-accuracy multitemporal laser scanning and flow data. *Water Resour. Res.* **2014**, *50*, 5532–5559. [CrossRef]

75. Rusnák, M.; Lehotský, M.; Kidová, A. Channel migration inferred from aerial photographs, its timing and environmental consequences as responses to floods: A case study of the meandering Topl'a River, Slovak Carpathians. *Morav. Geogr. Rep.* **2016**, *24*, 32–43. [CrossRef]

76. Tosic, R.; Lovric, N.; Dragicevic, S. Land use changes caused by bank erosion along the lower part of the Bosna river from 2001 to 2013. *Glas. Srp. Geogr. Drus. (Bull. Serbian Geogr. Soc)*. **2014**, *94*, 49–58. [CrossRef]

77. Das, T.K.; Haldar, S.K.; Sarkar, D.; Borderon, M.; Kienberger, S.; Das Gupta, I.; Kundu, S.; Guha-Sapir, D. Impact of riverbank erosion: A case study. *Australas. J. Disaster Trauma Stud.* **2017**, *21*, 73–81.

78. Szalai, Z.; Balogh, J.; Jakab, G. Riverbank erosion in Hungary—with an outlook on environmental consequences. *Hung. Geogr. Bull.* **2013**, *62*, 233–245.

79. Rahman, M.M.; Islam, M.N. Biodiversity Loss by Riverbank Erosion: A Study on the two Char Unions in Bangladesh. *J. Biodivers. Endanger. Species* **2018**, *6*, 1–6. [CrossRef]

80. Szabó, S.; Bertalan, L.; Kerekes, Á.; Novák, T.J. Possibilities of land use change analysis in a mountainous rural area: A methodological approach. *Int. J. Geogr. Inf. Sci.* **2015**, *30*, 708–726. [CrossRef]

81. Hunter, M.L.; Hunter, M.L., Jr. *Maintaining Biodiversity in Forest Ecosystems*; Cambrige University Press: Cambrige, UK, 1999.

82. Vizslán, T.; Szentgyörgyi, P. A Sajó-Hernád sík és a Sajó-völgy gerinces faunájáról (Vertebral Fauna of the Sajó-Hernád Plain and the Sajó Valley). *Fol. Hist.-Nat. Mus. Matr.* **1992**, *17*, 199–208.

83. Szép, T.; Nagy, K.; Nagy, Z.; Halmos, G. Population trends of common breeding and wintering birds in hungary, decline of long-distance migrant and farmland birds during 1999–2012. *Ornis Hung.* **2012**, *20*, 13–63. [CrossRef]

84. Girvetz, E.H. Removing erosion control projects increases bank swallow (*Riparia riparia*) population viability modeled along the Sacramento River, California, USA. *Biol. Conserv.* **2010**, *143*, 828–838. [CrossRef]

85. Szép, T. Partifecske (*Riparia riparia*). In *Birds of Hungary*; Haraszthy, L., Ed.; Mezőgazda Kiadó: Budapest, Hungary, 2000.

86. Szép, T.; Møller, A.P. Cost of parasitism and host immune defence in the sand martin Riparia riparia: A role for parent-offspring conflict? *Oecologia* **1999**, *119*, 9–15. [CrossRef] [PubMed]

87. Keesstra, S.; Nunes, J.P.; Saco, P.; Parsons, T.; Poeppl, R.; Masselink, R.; Cerdà, A. The way forward: Can connectivity be useful to design better measuring and modelling schemes for water and sediment dynamics? *Sci. Total Environ.* **2018**, *644*, 1557–1572. [CrossRef]

88. López-Vicente, M.; Quijano, L.; Palazón, L.; Gaspar, L.; Navas, A. Assessment of soil redistribution at catchment scale by coupling a soil erosion model and a sediment connectivity index (central spanish pre-pyrenees). *Cuad. Investig. Geográfica* **2015**, *41*, 127. [CrossRef]

89. Kavian, A.; Mohammadi, M.; Gholami, L.; Rodrigo-Comino, J. Assessment of the Spatiotemporal Effects of Land Use Changes on Runoff and Nitrate Loads in the Talar River. *Water* **2018**, *10*, 445. [CrossRef]

90. Lehotský, M.; Rusnák, M.; Kidová, A.; Dudžák, J. Multitemporal assessment of coarse sediment connectivity along a braided-wandering river. *L. Degrad. Dev.* **2017**, *29*, 1249–1261. [CrossRef]

91. Cossart, É.; Fressard, M. Assessment of structural sediment connectivity within catchments: Insights from graph theory. *Earth Surf. Dyn.* **2017**, *5*, 253–268. [CrossRef]

92. Hou, W.; Neubert, M.; Walz, U. A simplified econet model for mapping and evaluating structural connectivity with particular attention of ecotones, small habitats, and barriers. *Landsc. Urban Plan.* **2017**, *160*, 28–37. [CrossRef]

93. Basatnia, N.; Hossein, S.A.; Rodrigo-Comino, J.; Khaledian, Y.; Brevik, E.C.; Aitkenhead-Peterson, J.; Natesan, U. Assessment of temporal and spatial water quality in international Gomishan Lagoon, Iran, using multivariate analysis. *Environ. Monit. Assess.* **2018**, *190*, 314. [CrossRef] [PubMed]

94. Yu, D.; Liu, Y.; Xun, B.; Shao, H. Measuring Landscape Connectivity in a Urban Area for Biological Conservation. *CLEAN—Soil Air Water* **2013**, *43*, 605–613. [CrossRef]

Article

Effects of Different Spatial Configuration Units for the Spatial Optimization of Watershed Best Management Practice Scenarios

Liang-Jun Zhu [1,2], Cheng-Zhi Qin [1,2,3,*], A-Xing Zhu [1,2,3,4,5], Junzhi Liu [3,4] and Hui Wu [6]

[1] State Key Laboratory of Resources and Environmental Information System, Institute of Geographic Sciences and Natural Resources Research, CAS, Beijing 100101, China; zlj@lreis.ac.cn (L.-J.Z.); azhu@wisc.edu (A-X.Z.)

[2] University of Chinese Academy of Sciences, Beijing 100049, China

[3] Jiangsu Center for Collaborative Innovation in Geographical Information Resource Development and Application, Nanjing 210023, China; liujunzhi@njnu.edu.cn

[4] Key Laboratory of Virtual Geographic Environment (Nanjing Normal University), Ministry of Education, Nanjing 210023, China

[5] Department of Geography, University of Wisconsin-Madison, Madison, WI 53706, USA

[6] Smart City Research Center, Hangzhou Dianzi University, Hangzhou 310012, China; wuhui@hdu.edu.cn

* Correspondence: qincz@lreis.ac.cn; Tel.: +86-10-648-889-59

Received: 22 December 2018; Accepted: 31 January 2019; Published: 2 February 2019

Abstract: Different spatial configurations (or scenarios) of multiple best management practices (BMPs) at the watershed scale may have significantly different environmental effectiveness, economic efficiency, and practicality for integrated watershed management. Several types of spatial configuration units, which have resulted from the spatial discretization of a watershed at different levels and used to allocate BMPs spatially to form an individual BMP scenario, have been proposed for BMP scenarios optimization, such as the hydrologic response unit (HRU) etc. However, a comparison among the main types of spatial configuration units for BMP scenarios optimization based on the same one watershed model for an area is still lacking. This paper investigated and compared the effects of four main types of spatial configuration units for BMP scenarios optimization, i.e., HRUs, spatially explicit HRUs, hydrologically connected fields, and slope position units (i.e., landform positions at hillslope scale). The BMP scenarios optimization was conducted based on a fully distributed watershed modeling framework named the Spatially Explicit Integrated Modeling System (SEIMS) and an intelligent optimization algorithm (i.e., NSGA-II, short for Non-dominated Sorting Genetic Algorithm II). Different kinds of expert knowledge were considered during the BMP scenarios optimization, including without any knowledge used, using knowledge on suitable landuse types/slope positions of individual BMPs, knowledge of upstream–downstream relationships, and knowledge on the spatial relationships between BMPs and spatial positions along the hillslope. The results showed that the more expert knowledge considered, the better the comprehensive cost-effectiveness and practicality of the optimized BMP scenarios, and the better the optimizing efficiency. Thus, the spatial configuration units that support the representation of expert knowledge on the spatial relationships between BMPs and spatial positions (i.e., hydrologically connected fields and slope position units) are considered to be the most effective spatial configuration units for BMP scenarios optimization, especially when slope position units are adopted together with knowledge on the spatial relationships between BMPs and slope positions along a hillslope.

Keywords: spatial configuration units; best management practices (BMPs); spatial optimization; hydrologic response units (HRUs); hydrologically connected fields; slope positions; watershed process simulation

---

1. Introduction

Different best management practices (or beneficial management practices, BMPs for short) scenarios (i.e., spatial configurations of multiple BMPs) at the watershed scale may have significantly different environmental effectiveness, economic efficiency, and practicality [1–6]. They are valuable for decision-making of integrated watershed management to assess the environmental effectiveness and economic efficiency of watershed BMP scenarios and then propose optimal ones. Currently, a popular approach to achieving this target is based on watershed modeling [7–9] coupled with intelligent optimization algorithms [2,4,5,10–13], so-called BMP scenarios optimization. To conduct the BMP scenarios optimization, each individual BMP scenario is created by automatically selecting and allocating BMPs on spatial configuration units (also called BMP configuration units hereafter), which have resulted from the spatial discretization of a watershed at one among different levels (such as subbasins, and hydrologic response units; Figure 1). When a specific type of BMP configuration unit is chosen for the BMP scenarios optimization of a watershed, normally each individual BMP configuration unit in the watershed is allowed to be configured with only one type of BMP (as the situation in this study). Then the effects of the scenario on watershed behavior are simulated by watershed models [4,13]. The simulation result is the basis of the automatic spatial optimization of BMP scenarios. Therefore, the determination of the BMP configuration units becomes the one key issue for the BMP scenarios optimization.

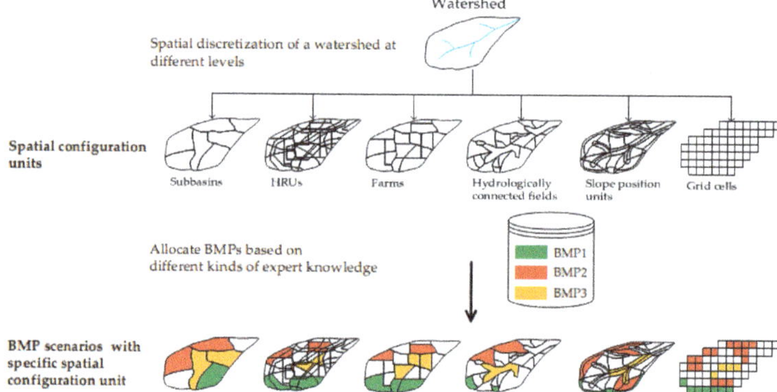

Figure 1. Schematic diagram of spatial discretization of a watershed at different levels (such as sub-basins, hydrologic response units (HRUs), farms, hydrologically connected fields, slope position units, and grid cells) and the corresponding best management practice (BMP) scenario examples after allocating multiple types of BMPs based on different kinds of expert knowledge.

Currently, spatial configuration units used in BMP scenarios optimization mainly include five types with different levels, i.e., subbasins [14–17], hydrologic response units (HRUs) [12,18,19], farms [10,11], hydrologically connected fields [5], and slope position units (i.e., landform positions at hillslope scale, such as ridge, backslope, etc.) [2].

Subbasins are normally regarded as relatively closed and independent spatial configuration units which can be further delineated into different levels of spatial configuration units such as landform positions, HRUs, and gridded cells [20]. Thus, subbasins are coarse-grained units for configuring spatially explicit BMPs, and it is suitable to serve as BMP configuration units for situations in which the potential locations of BMPs are predefined within each subbasin [17]. Meanwhile, using subbasins as BMP configuration units can decrease the search space of spatial optimization more than that of the adoption of other more detailed spatial configuration units, and thus can achieve a better optimizing efficiency [15]. However, using subbasins as BMP configuration units is not available for configuring multiple spatially explicit BMPs within one subbasin while BMPs often have different effects on different spatial positions of a hillslope [2,21,22].

HRUs are delineated as hydrologic homogeneous areas according to landuse, soil, and topography (commonly presented through slope percentage) within one subbasin [23]. HRUs are extensively used for the sake of convenience, especially when the Soil and Water Assessment Tool (SWAT) [23] is applied to watershed modeling. However, an HRU is spatially discrete—it may occupy the hillslope from ridge to valley bottom—and is even spatially ambiguous when the three thresholds introduced in SWAT for delineating HRUs (i.e., the minimum percentage of the landuse area over the subbasin area, the soil area over the landuse area, and the slope class area over the soil area) are set as non-zero values [24]. Although setting these thresholds to be non-zero can reduce the count of HRUs and hence improve the computational efficiency of simulations, the inappropriate representation of the study area may introduce some levels of ambiguity in simulations. Therefore, the selected optimal BMP scenarios may not be easy to implement spatially explicitly. Similar to subbasins, HRUs are also inappropriate to serve as BMP configuration units for spatially explicit BMPs. Furthermore, the impact of BMPs of an upslope HRU on a downslope HRU cannot be represented [24], although this impact is important for incorporating the expert knowledge of spatial relationships between BMPs and spatial positions [2,5].

To improve the practicality of BMP scenarios, some studies adopted farms [10,11], or the so-called spatially explicit HRUs that are directly defined by farm boundaries [25] or landuse/landcover field maps [26], as BMP configuration units. These BMP configuration units are spatially one-to-one matched with farm or landuse/landcover fields and thus can be collectively referred to as spatially explicit HRUs [26]. Compared with spatially discrete HRUs, spatially explicit HRUs can make corresponding BMP scenarios more practical to stakeholders (such as farmers, landowners, or land managers). Meanwhile, due to the ignorance of topographic variance inside each farm or landuse/landcover field, such BMP configuration units normally have a smaller count than spatially discrete HRUs for a study area, which means higher optimizing efficiency [1,26]. However, there are still no explicitly defined upstream–downstream relationships between spatially explicit HRUs, which means that the spatial relationships between BMPs and spatial positions cannot be represented effectively [2,5].

Wu et al. [5] proposed hydrologically connected fields with upstream–downstream relationships to be BMP configuration units, which can be delineated by considering spatial topology based on flow directions and a landuse map. With hydrologically connected fields, a set of expert rules of BMP interactions based on the upstream–downstream relationships can be developed in the form of "if-then" rules, i.e., if a field has been configured with one BMP, its adjacent upstream fields should not be configured with BMP, otherwise, BMP will be randomly selected and configured on the adjacent upstream fields according to their landuse type [5]. To the best of our knowledge, the study of Wu et al. [5] is the first work on incorporating the spatial relationships between BMPs and spatial positions into BMP scenarios optimization, so that its cost-effectiveness and optimizing efficiency could be improved. However, hydrologically connected fields may be delineated across multiple landform positions or subbasins, since the hydrological relationship, considered by Wu et al. [5], is built at the watershed scale rather than the hillslope or subbasin scale. This means that hydrologically connected fields have weak spatial relationships to homogeneous functional units at the hillslope scale from the perspectives of physical geography such as geomorphic, soil, and hydrologic conditions [2]. Therefore, hydrologically connected fields also face the shortcoming that the spatially explicit BMPs cannot be represented effectively. A possible way for spatially explicit HRUs and hydrologically connected fields to overcome the above-mentioned shortcoming is to delineate them so as to be small enough patchworks of gridded cells within homogeneous functional units, or even individual gridded cells [27–29]. However, this approach will render the optimization based on such spatial configuration units computationally intensive or even unsolvable [27]. This makes such an approach only suitable for the spatial optimization of one single BMP within a little watershed [28,29], thus having too narrow applicability for normal watershed management which considers multiple BMPs.

More recently, Qin et al. [2] proposed slope position units [30,31] as BMP configuration units, so as to overcome the above-mentioned shortcomings of hydrologically connected fields and other BMP configuration units. Slope position units (e.g., ridge, backslope, and valley), as spatially contiguous and topographically connected units along the hillslope, are inherently related to physical watershed processes [20,30,32], and thus affect the effectiveness of BMPs [2]. Besides, the count of slope position units is comparatively limited, so as to ensure optimizing efficiency. Therefore, the spatial relationships between BMPs and slope positions along the hillslope could be explicitly and effectively considered during the spatial optimization of BMP scenarios. According to the preliminary study [2], this spatial-relationship-considered way of using slope position units as BMP configuration units is effective, efficient, and practical for BMP scenarios optimization, compared to a standard random optimization way of selecting and allocating BMPs randomly on configuration units.

Although many studies have assessed the cost-effectiveness of each individual type of above-mentioned BMP configuration units for the spatial optimization of BMP scenarios, as far as we know, a comparison among main types of BMP configuration units based on the same one watershed model for an area is still lacking. To discuss the effects of different BMP configuration units for watershed BMP scenarios optimization with regard to cost-effectiveness, optimizing efficiency, and practicality, this paper compares four types of BMP configuration units (i.e., HRUs, spatially explicit HRUs, hydrologically connected fields, and slope position units) in the spatial optimization of configuring multiple spatially explicit BMPs for mitigating soil erosion based on a spatially distributed watershed model. Subbasins are not included in this study, because subbasins are too coarse-grained to represent multiple spatially explicit BMPs.

2. Materials and Methods

2.1. Methodology

To compare the effects of these four types of BMP configuration units for watershed BMP scenarios optimization, a widely used spatial optimization framework of BMP scenarios based on watershed modeling coupled with intelligent optimization algorithms [2,10,12] was adopted in this study. As shown in Figure 2, the spatial optimization framework of watershed BMP scenarios mainly consists of four components: (1) BMP configuration units for allocating BMPs within the watershed; (2) a BMP knowledge base together with BMP configuration units as inputs to generate and evaluate BMP scenarios; (3) models for evaluating each watershed BMP scenario, including a distributed watershed model that can simulate spatial interactions between spatially explicitly distributed BMPs and effectively assess the environmental effectiveness of each BMP scenario [2,5,33], and a BMP scenario cost model for estimating the economic efficiency of each BMP scenario; (4) a multi-objective optimization component based on an intelligent optimization algorithm such as NSGA-II (Non-dominated Sorting Genetic Algorithm II) [34], which includes initializing BMP scenarios based on BMP configuration units and a BMP knowledge base, and generating new BMP scenarios or proposing optimal ones based on the evaluation results of all current BMP scenarios. These components are elaborated in four subsections (Sections 2.3–2.6) followed by the subsection of the study area and dataset (Section 2.2).

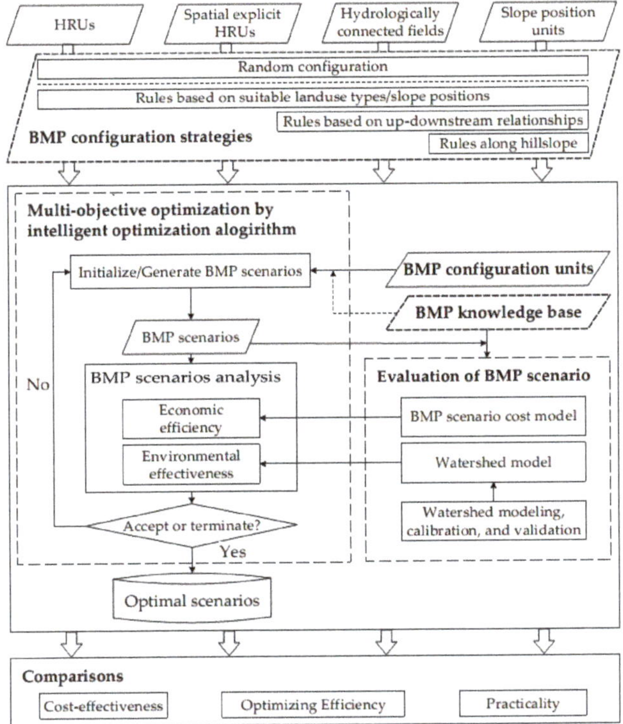

Figure 2. Framework comparing the effects of different BMP configuration units for the spatial optimization of watershed BMP scenarios in this study (extended from [2]).

Besides knowledge on the environmental effectiveness and cost-benefit of individual BMPs, BMP configuration knowledge (such as the suitable landuse types/slope positions for individual BMPs, and the spatial relationships between BMPs and spatial positions) is necessary for BMP scenarios optimization (Figure 2). According to the characteristics of BMP configuration units, different kinds of BMP configuration knowledge can be saved in the BMP knowledge base and applied (hereafter referred to as BMP configuration strategy) (Figure 2). For example, the simple rule of suitable landuse types/slope positions is applicable for all types of BMP configuration units, while the expert rules based on upstream–downstream relationships between BMPs can only be applied to the BMP configuration units which have upstream–downstream relationships [5], such as hydrologically connected fields and slope position units. Furthermore, the expert rules based on the spatial relationships between BMPs and slope positions along the hillslope are only applicable for slope position units [2]. Section 2.4 contains more detailed information about the BMP knowledge base and BMP configuration strategies adopted in this study.

To assess the effects of four types of BMP configuration units for watershed BMP scenarios optimization, all feasible combinations of BMP configuration units and BMP configuration strategies were investigated by the same watershed BMP scenarios optimization framework (Figure 2). The results of different BMP configuration units applied with available BMP configuration strategies are discussed from the perspectives of cost-effectiveness, optimizing efficiency, and practicality (Figure 2). The detailed experimental design is described in Section 2.7.

2.2. Study Area and Dataset

The Youwuzhen watershed (~5.39 km^2), located in the typical severely eroded red-soil hilly region in southeastern China, was selected as the study area (Figure 3). Low hills with steep slopes (up to 52.9° and with an average slope of 16.8°) and broad alluvial valleys are the primary geomorphology forms [2]. Forest (59.8%), paddy field (20.6%), and orchard (12.8%) are the primary landuse types in the study area (Figure 4). Additionally, forests in the study area are dominated by secondary or human-made forests with low coverage due to the destruction of vegetation caused by soil erosion and economic development in the past [35] (Figure 4). Soil types in the study area are red soil (78.4%) and paddy soil (21.6%) which can be classified as Ultisols and Inceptisols in US Soil Taxonomy, respectively [36]. Red soil is mainly distributed in the hilly region, while paddy soil is mainly distributed in the broad alluvial valleys with a similar spatial pattern of paddy rice landuse (Figure 4). The climate belongs to the mid-subtropical monsoon moist climate with an annual average temperature of 18.3 °C. The annual average precipitation is 1697.0 mm [35].

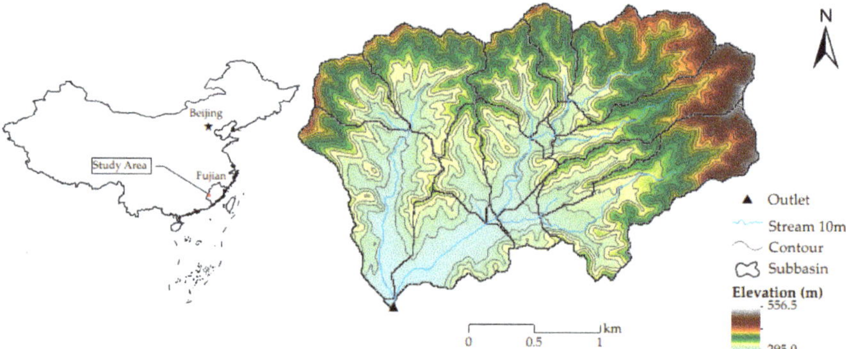

Figure 3. Map of the Youwuzhen watershed in Fujian Province, China (adapted from [2]).

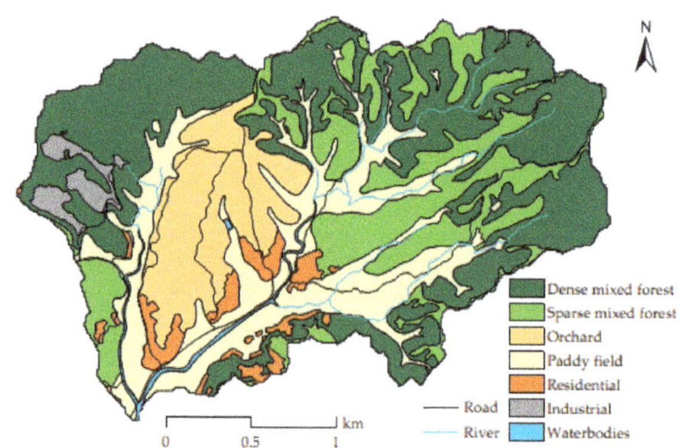

Figure 4. Map of landuse in the study area.

The basic spatial data collected for watershed modeling of the Youwuzhen watershed include a gridded Digital Elevation Model, a soil type map, and a landuse type map, which were all unified to be of 10 m resolution [2]. Soil properties were derived from field sampling data [35]. Landuse/landcover related parameters were referenced from the SWAT database [37] and relevant literature [38]. The climate data containing daily meteorological data and precipitation data from 2012 to 2015 were derived from the National Meteorological Information Center of China Meteorological Administration and local monitoring stations, respectively. The periodic site-monitoring streamflow and sediment discharge data of the watershed outlet from 2012 to 2015 were provided by the Soil and Water Conservation Bureau of Changting County, Fujian province, China.

With an accumulated threshold of 0.185 km^2 for the study area [35], the Youwuzhen watershed was delineated into 17 subbasins (Figure 3). The streamflow and sediment discharge data were screened by a rule that requires complete records of rainstorms with more than three consecutive days for watershed modeling due to the limited data quality [2]. Finally, the year 2012 was selected as a warm-up period for watershed modeling, the years 2014 and 2015 for calibration, and the year 2013 for validation.

2.3. Delineation of BMP Configuration Units

2.3.1. HRUs

The QSWAT, an open source user interface for the SWAT model [39], was used to delineate the typical spatially discrete HRUs by overlaying the landuse map, soil map, and the classification map of the slope percentage. The slope percentage was classified into five classes with nearly equal areas according to the quantile classification method, i.e., 0%–9%, 9%–23%, 23–36%, 36%–48%, and larger than 48%. Three threshold values introduced in SWAT to delineate HRUs were all set to 0%, so as to obtain a full spatial coverage of HRUs (Figure 5a). Finally, a total of 355 HRUs were generated in the study area (Figure 5a).

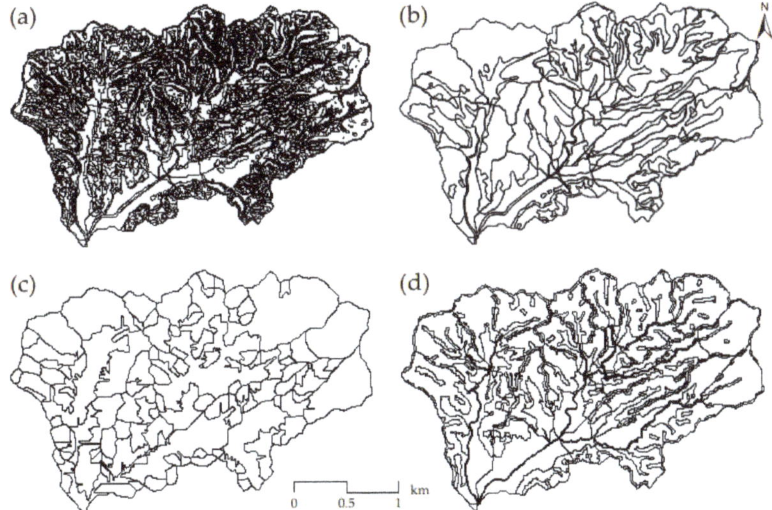

Figure 5. Delineations of BMP configuration units of the Youwuzhen watershed: (**a**) HRUs; (**b**) spatially explicit HRUs; (**c**) hydrologically connected fields; (**d**) slope position units.

2.3.2. Spatially Explicit HRUs

The method proposed by Teshager et al. [26] was adopted to delineate the spatially explicit HRUs. First, the landuse map was split by river and road networks—not forest type—in the map. Then, the split landuse map was intersected by the subbasin boundary. After eliminating small polygons, landuse polygons were re-labeled by assigning different codes for the polygons with the same landuse type within a subbasin. For example, three polygons with a landuse of orchard (ORCD) within a subbasin were re-assigned to ORCD1, ORCD2, and ORCD3, respectively. Similarly, the processed landuse map was intersected by the soil map and the landuse-soil polygons with the same soil type within a subbasin were re-labeled according to soil types. A single slope percentage class was implicitly used in this method to avoid HRU fragmentation and ensure that each HRU matched individual fields in a subbasin. Finally, 166 spatially explicit HRUs were generated (Figure 5b).

2.3.3. Hydrologically Connected Fields

The basic idea of the delineation algorithm of the hydrologically connected field proposed by Wu et al. [5] is to build a gridded cell tree structure based on flow directions and then merge gridded cells with the same landuse type in this tree structure into one field. A threshold value of the minimum size of a field was set to eliminate tiny fields by aggregating them into their downslope large ones. The smaller the threshold value, the more the fields will be delineated. The threshold value was insensitive to the cost-effectiveness of the optimization of BMP scenarios according to the sensitivity analysis conducted by Wu et al. [5]. Therefore, under the consideration of optimizing efficiency and comparability with other BMP configuration units, 103 hydrologically connected fields were delineated with a threshold value of 70 cells (Figure 5c).

2.3.4. Slope Position Units

The same as in the case study of Qin et al. [2], a simple system of three types of slope positions (i.e., ridge, backslope, and valley) was adopted to delineate slope position units. The automated program developed by Zhu et al. [31] was used to derive fuzzy memberships of gridded cells to each slope position and then a crisp classification map of slope positions was generated by the maximization principle [30]. The slope position map was then intersected by the hillslope boundary to ensure each hillslope had a full sequence of slope position units from the top to the bottom of the hillslope. Finally, 105 slope position units within 35 hillslopes were delineated.

2.4. BMP Knowledge Base and BMP Configuration Strategies

As shown in Table 1, four BMPs that had been widely implemented in Changting County for ecological restoration and soil and water conservation were selected in this study, i.e., Closing measures (CM), Arbor-bush-herb mixed plantation (ABHMP), Low-quality forest improvement (LQFI), and Orchard improvement (OI) [2,40].

The BMP knowledge base used in this study mainly includes two categories of knowledge, i.e., the environmental effectiveness and cost-benefit knowledge used for evaluating BMP scenarios, and the BMP configuration knowledge used with BMP configuration strategies. The four BMPs considered in this study can improve soil properties through long- or/and short-term processes and thus achieve environmental effectiveness (i.e., improving water conservation and mitigating soil erosion) [39]. Therefore, the environmental effectiveness of the four BMPs considered in this study can be mainly represented by the improvements of soil properties and the change of USLE (Universal Soil Loss Equation) factors in the area with these BMPs in the watershed model [2,5,8]. The Fujian Soil and Water Conservation Monitoring Station monitored the dynamic changes in the soil properties such as organic matter, bulk density, and total porosity improved by various BMPs by taking samples annually from 2000 to 2008 [41]. This study assumed that the long-term environmental effectiveness of BMPs can reach relative stability after several years of maintenance from their first establishment [42]. Therefore, the relative changes of the 8-year sampled and derived soil properties (Table 2) were used to represent the environmental effectiveness of BMPs in the watershed modeling. Besides, the relative changes of the conservation practice factor of USLE (i.e., USLE_P) in Table 2 were adopted from the calibrated SWAT model for this area [35]. The BMP cost-benefit data including initial implementation cost, annual maintenance cost, and annual benefit, were estimated from reports of local government projects [2] (Table 2).

Table 1. Brief descriptions of the four BMPs considered in this study (adapted from [2,40]; photos from [35]).

BMP	Photo	Brief Description
Closing measures (CM)		Closing the ridge area and/or upslope positions from human disturbance (e.g., ban on felling tree and grazing) to facilitate afforestation.
Arbor–bush–herb mixed plantation (ABHMP)		Planting trees (e.g., *Schima superba* and *Liquidambar formosana*), bushes (e.g., *Lespedeza bicolor*), and herbs (e.g., *Paspalum wettsteinii*) in level trenches with compound fertilizer on hillslopes.
Low-quality forest improvement (LQFI)		Improving the infertile forest located in the upslope and steep backslope positions by applying compound fertilizer to the hole with a size of 40 cm × 40 cm × 40 cm in the uphill position of crown projection.
Orchard improvement (OI)		Improving orchards on the middle and down slope positions under better water and fertilizer conditions by constructing level terraces, drainage ditches, storage ditches, irrigation facilities, and roads, planting economic fruit, and interplanting grasses and Fabaceae (*Leguminosae*) plants.

BMP configuration knowledge is normally in the form of rules. They include knowledge on individual BMPs (such as its suitable landuse types, suitable slope positions, and overall environmental effectiveness grade; Table 3) according to the characteristics of individual BMPs [35,41] (Table 1), and knowledge on spatial relationships between BMPs (such as the upstream–downstream relationships between BMP configuration units, see below) [2,5]. The overall environmental effectiveness grade of a BMP ranges from 1 to 5 which represents the improvement degree of mitigating soil erosion for an area with the BMP (the higher the better; Table 3). The overall environmental effectiveness grade can be used to formalize the expert knowledge on the spatial configuration of BMPs along the hillslope scale [2].

Table 2. Environmental effectiveness and cost-benefit knowledge of the four BMPs (CM = Closing measures, ABHMP = Arbor–bush–herb mixed plantation, LQFI = Low-quality forest improvement, OI = Orchard improvement) (adapted from [2]).

BMP	Environmental Effectiveness [1]						Cost-Benefit (CNY 10,000/km^2)		
	OM	BD	PORO	SOL_K	USLE_K	USLE_P	Initial	Maintain/yr	Benefit/yr
CM	1.22	0.98	1.02	0.81	1.01	0.90	15.5	1.5	2.0
ABHMP	1.45	0.93	1.07	1.81	0.82	0.50	87.5	1.5	6.9
LQFI	1.05	0.87	1.13	1.71	1.71	0.50	45.5	1.5	3.9
OI	2.05	0.96	1.03	1.63	1.63	0.75	420	20	60.3

[1] Environmental effectiveness of BMPs includes soil property parameters (i.e., OM = organic matter, BD = bulk density, PORO = total porosity, and SOL_K = soil hydraulic conductivity) and universal soil loss equation (USLE) factors (i.e., USLE_K = soil erodibility factor and USLE_P = conservation practice factor). Values in this column represent relative changes (i.e., multiplying) to the original property values and thus have no units.

Table 3. Suitable landuse types/slope positions and the overall environmental effectiveness grade of the four BMPs (CM = Closing measures, ABHMP = Arbor–bush–herb mixed plantation, LQFI = Low-quality forest improvement, OI = Orchard improvement) (adapted from [2]).

BMP	Suitable Landuse Types	Suitable Slope Positions	Effectiveness Grade
CM	forest	ridge, backslope	3
ABHMP	forest, orchard	ridge, backslope, and valley	5
LQFI	forest	backslope	4
OI	forest, orchard	valley	4

According to knowledge (or rules) used for BMP configuration, four BMP configuration strategies were adopted during assessment of the effects of four types of BMP configuration units for BMP scenarios optimization (Figure 2).

1. Random configuration strategy (RAND for short). Without knowledge used, the RAND strategy randomly selects and allocates one of the four BMPs on the BMP configuration units. Thus, it can be used for any type of BMP configuration unit considered in this study.

2. Strategy with knowledge on the suitable landuse types/slope positions of individual BMPs (SUIT for short). The SUIT strategy is to randomly select and allocate one of the suitable BMPs according to its suitable landuse type and/or slope position to the BMP configuration unit. This strategy is applicable for any BMP configuration units with landuse type and/or slope position type.

3. Strategy based on expert knowledge of upstream–downstream relationships [5] (UPDOWN for short). Extended from the SUIT strategy, the UPDOWN strategy applies the expert rules of BMP interactions ("if–then" rules) based on the upstream–downstream relationships between BMP configuration units to generating BMP scenarios. That is, if a field has been configured with one BMP, its adjacent upstream fields should not be configured with a BMP; otherwise the BMP will be randomly selected and configured on the adjacent upstream fields according to their landuse types. This strategy is available for hydrologically connected fields and slope position units.

4. Strategy with expert knowledge on the spatial relationships between BMPs and slope positions along the hillslope [2] (HILLSLP for short). The HILLSLP strategy further extends the SUIT strategy to consider spatial constraints among BMPs on different slope positions along the hillslope from upstream to downstream. Such expert knowledge used in this study was adapted from Qin et al. [2], i.e., the effectiveness grade of the BMP configured on the downstream unit of a hillslope should be greater than or equal to that of the BMP configured on the upstream unit of the same hillslope. For example, according to the knowledge in Table 3, if the backslope unit has been configured with ABHMP, the downstream valley unit of the same hillslope may be configured with ABHMP or without BMP, while the upstream ridge unit has three configuration options (i.e., CM, ABHMP, and without BMP).

2.5. Watershed Model and BMP Scenario Cost Model

Note that BMP configuration units are not necessary to be consistent with the basic simulation unit (e.g., gridded cell) of watershed models. To simulate the spatial interactions between spatially explicitly distributed BMPs effectively, a fully distributed watershed model was constructed based on an open-source, modular, and parallelized watershed modeling framework, Spatially Explicit Integrated Modeling System (SEIMS, https://github.com/lreis2415/SEIMS; [33]) in this study. With the flexible modular structure and the parallel-computing middleware [33,43], SEIMS allows users to add their own algorithms in a nearly serial programming manner and to customize parallelized watershed models according to the characteristics of the study area and the application requirements. SEIMS also supports model-level parallel computation for applications which need numerous model runs, such as BMP scenarios optimization in this study.

To evaluate the long-term effects of BMP scenarios on mitigating soil erosion, a SEIMS-based watershed model with gridded cells as the basic simulation unit was constructed to simulate hydrology, soil erosion, plant growth, and nutrient cycling processes at a daily time-step [2]. The hydrology processes considered in this watershed model include interception, surface depression storage, surface runoff, potential evapotranspiration, percolation, interflow, groundwater flow, and channel flow. Soil erosion on hillslopes was estimated by the Modified Universal Soil Loss Equation (MUSLE) [44], and sediment routing in channels was simulated by a simplified Bagnold stream power equation adapted in the SWAT model. The plant growth process and nutrient (i.e., nitrogen and phosphorous) cycling process were also adapted from SWAT. More details about the corresponding simulation algorithms can be found in Qin et al. [2].

Note that the BMP module of SEIMS applies BMP knowledge to update the input parameters on each of the gridded cells configured with BMP suitable for current landuse on the cell before the simulation, while the non-effective configurations are ignored. The non-effective configurations may occur in two circumstances: (1) the BMP configuration unit includes multiple landuse types (e.g., the generalized hydrologically connected fields and slope position units) and some of them with small areas are unsuitable for the currently configured BMP; (2) the BMP configuration unit applied with the random configuration strategy, e.g., the BMP configuration unit with paddy rice landuse configured with Low-quality forest improvement (LQFI) or Orchard improvement (OI) practices.

After conducting the parameter sensitivity analysis using the Morris screening method [45], the SEIMS-based watershed model in the study area was calibrated by an auto-calibration procedure based on the NSGA-II algorithm (a multi-objective optimization algorithm extended from the Genetic Algorithm; [34]) provided in SEIMS [33]. The Morris screening method is a so-called one-step-at-a-time (OAT) global sensitivity analysis method, which was used to qualitatively identify important parameters for the simulation of streamflow and sediment export at the outlet in this study. Then a small number of sensitivity parameters (such as the baseflow exponent and the baseflow recession coefficient for groundwater and soil water capacity) were selected for auto-calibration in this study [33]. The NSGA-II was also applied to BMP scenarios optimization in this study, and is described in detail in Section 2.6. One optimal calibration solution was selected as the baseline scenario (Table 4). The calibration and validation results of streamflow ($m^3 s^{-1}$) and sediment export (kg) at the outlet of Youwuzhen watershed were evaluated by widely-used model performance indicators such as NSE (Nash–Sutcliffe Efficiency), PBIAS (Percent BIAS), and RSR (Root mean Square error–standard deviation Ratio) [46]. According to the general performance ratings for simulations at a monthly time step [46], the model performance is satisfactory when NSE ≥ 0.50, RSR ≤ 0.70, and the absolute value of PBIAS $\leq 25\%$ (55% for sediment). Thus, the streamflow performance is nearly satisfactory, while the sediment is relatively poor because of a few unreasonable peak values (Table 4). Nevertheless, the general trends of hydrographs in the study area can be captured by the calibrated SEIMS-based model according to visual judgement. Therefore, it is acceptable to apply the calibrated SEIMS-based watershed model to BMP scenarios optimization. The annual average sediment yields from 2013 to

2015 of the entire watershed, under each BMP scenario for spatial optimization, were calculated by this model.

Table 4. The calibration (2014–2015) and validation (2013) results of the baseline scenario by the Spatially Explicit Integrated Modeling System (SEIMS)-based watershed model. (NSE = Nash–Sutcliffe Efficiency, RSR = Root mean Square error–standard deviation Ratio, PBIAS = Percent BIAS).

Constituent	Calibration Period (2014–2015)			Validation Period (2013)		
	NSE	RSR	PBIAS (%)	NSE	RSR	PBIAS (%)
Streamflow ($m^3 s^{-1}$)	0.50	0.71	13.55	0.57	0.65	−14.71
Sediment export (kg)	0.30	0.84	13.93	0.45	0.74	−42.39

Besides the SEIMS-based watershed model for evaluating environmental effectiveness, a simple BMP scenario cost model (Equation (1)) was adopted to calculate the net cost of each BMP scenario according to the cost-benefit knowledge in the BMP knowledge base [2].

$$f_{\text{net}-\text{cost}}(X) = \sum_{i=1}^{n} A(x_i) \times \{[C(x_i) + yr \times (M(x_i) - B(x_i))]\} \tag{1}$$

where $f_{\text{net-cost}}(X)$ is the net cost of a BMP scenario (represented as X); n is the count of BMP configuration units; $A(x_i)$ is the area covered by the BMPs implemented in the i th configuration unit; yr is the years when the effectiveness of BMPs reach stability, which is 8 in this study (see Section 2.4); $C(x_i)$, $M(x_i)$, and $B(x_i)$ are unit costs (CNY 10,000/km^2) for initial implementation, annual maintenance, and annual benefit (Table 2), respectively.

2.6. Multi-Objective Optimization by an Intelligent Optimization Algorithm

The objectives in this study are to minimize the net cost of the BMP scenario (Equation (2)) and maximize the reduction rate of soil erosion. The reduction rate of soil erosion of each BMP scenario is the relative change compared to the baseline scenario (Equation (3)).

$$optimal\ solutions = \min(-f_{\text{reduction}-\text{rate}}(X), f_{\text{net}-\text{cost}}(X)) \tag{2}$$

$$f_{\text{reduction}-\text{rate}}(X) = (v(0) - v(X))/v(0) \tag{3}$$

where $f_{\text{reduction-rate}}(X)$ is the reduction rate of soil erosion under the BMP scenario X compared to that under the baseline scenario; $v(0)$ and $v(X)$ are the total amount of soil erosion (kg) under the baseline scenario and the X scenario, respectively.

The NSGA-II [33], which has been successfully applied to many similar studies [2,16–18], was selected as the intelligent optimization algorithm in this study. As shown in Figure 2, the NSGA-II algorithm first initializes the initial BMP scenarios (called "population" in NSGA-II) based on one type of BMP configuration unit and one available BMP configuration strategy as described in Sections 2.3 and 2.4. A BMP scenario (called "individual") is represented as an array with a length equal to the number of BMP configuration units (called "chromosome") with the corresponding reduction rate of soil erosion and net cost values (called "fitness") to be evaluated. Each value of the chromosome (called "gene") stands for one selected BMP type or without BMP on a BMP configuration unit. Then, each individual of the initial population, as the initial "generation" for the following optimization, is evaluated by objective functions, i.e., the reduction rate of soil erosion is assessed by the calibrated watershed model and the net cost estimated by the BMP scenario cost model. The evaluated individuals are sorted, based on the non-domination of fitness, and selected by a specified number as an elite set for each generation which is known as near-optimal Pareto solutions [33]. Then a circular process of regenerating and evaluating BMP scenarios (i.e., new generation) proceeds until a user-assigned maximum generation number is reached. Individuals of current generation (called "offspring") combined with the near-optimal Pareto solutions of former generation (called "parent") are proceeded by non-dominated sorting, selection, crossover, and mutation operations to regenerate new BMP scenarios [33].

When a BMP configuration strategy with knowledge (i.e., SUIT strategy, UPDOWN strategy, or HILLSLP strategy) was adopted, it was incorporated into not only the initialization [5] but also the regeneration of BMP scenarios, i.e., crossover and mutation operations [2]. The crossover operation of the UPDOWN strategy proposed by Wu et al. [5] was extended as follows. The randomly selected BMP configuration unit (i.e., the position of the crossover gene) is first checked to see whether it can ensure that the two generated "children" individuals still conform to the "if–then" rules after the exchange of the subtrees in which the selected unit is the most downstream unit. If the current selected unit fails to meet the condition, the downstream units of the current one will be checked in order, until a qualified unit is reached. The finally reached unit, except for the most downstream unit of the entire study area, will be used as the crossover gene. Under the HILLSLP strategy, the BMPs are configured along each hillslope from bottom to top during the initialization, and the crossover operation is to exchange the randomly selected hillslopes (without breaking the configuration among all slope position units along the same hillslope). Such generated "children" individuals conform to the HILLSLP strategy. As for the mutation operation of the UPDOWN strategy and HILLSLP strategy, the potential BMP types for a BMP configuration unit (i.e., a mutant gene) are first determined according to the rule set of adopted knowledge and the BMP types of its upstream and downstream units (i.e., gene values). Then, a different BMP from the current one will be randomly selected for the mutant gene. In such a way, every BMP scenario generated during spatial optimization is reasonable in terms of the corresponding knowledge, and may result in higher optimizing efficiency.

2.7. Design of Comparison Experiment

In this study, four types of BMP configuration units (i.e., HRUs, spatially explicit HRUs (EXPLICITHRU), hydrologically connected fields (CONNFIELD), and slope position units (SLPPOS)) and four BMP configuration strategies (i.e., RAND strategy, SUIT strategy, UPDOWN strategy, and HILLSLP strategy) were combined according to their availability (see Section 2.4) for the spatial optimization of multiple spatially explicit BMPs. This means that, in total, 11 experiments of feasible combinations were conducted, i.e., HRU+RAND which means the combination of using HRU as BMP configuration units applied with the random configuration strategy (ex analogia), HRU+SUIT, EXPLICITHRU+RAND, EXPLICITHRU+SUIT, CONNFIELD+RAND, CONNFIELD+SUIT, CONNFIELD+UPDOWN, SLPPOS+RAND, SLPPOS+SUIT, SLPPOS+UPDOWN, and SLPPOS+HILLSLP. The Python script of the BMP scenarios optimization based on slope position units developed by Qin et al. [2] was extended to support multiple types of BMP configuration units and BMP configuration strategies considered in this study. The script built on the model-level parallel framework of SEIMS [32] distributes computing tasks dynamically across a Linux cluster to improve computation efficiency.

The parameter settings of the NSGA-II algorithm were kept the same for all experiments. The initial population size was 480 with a selection rate of 0.8 and a maximum generation number of 100 [5]. The crossover probability and the mutation probability were 0.8 and 0.1, respectively.

To explore the effects of different BMP configuration units for the spatial optimization of spatially explicit BMPs, the results of different BMP configuration units applied with the same BMP configuration knowledge are first compared. Then, the combinations of BMP configuration units and the corresponding optimal configuration strategies are compared. The optimal configuration strategy for a type of BMP configuration unit was considered to be the most reasonable one based on a BMP knowledge base and not necessarily the one with the best non-dominated Pareto solutions from the mathematical perspective. Thus, SLPPOS+HILLSLP, CONNFIELD+UPDOWN, EXPLICITHRU+SUIT, and HRU+SUIT were selected for this comparison according to the available level with the most BMP configuration knowledge for each BMP configuration unit (Section 2.4).

The results of different BMP configuration units will be discussed from the perspectives of cost-effectiveness, optimizing efficiency, and practicality. The near-optimal Pareto solutions plotted as a scatter plot can give a simple and direct interpretation of the convergence and diversity of different spatial optimization results. When the near-optimal Pareto solutions from different combinations are compared, the one with more non-dominated solutions indicates a better cost-effectiveness [33], which means more BMP scenarios with better multi-objective optimization provided as optimized solutions for decision-making. Besides, the hypervolume index [47], which measures the volume (or area for two-dimensions) of objective space covered by a set of near-optimal Pareto solutions, provides a quantitative comparison of the cost-effectiveness considering both convergence and diversity [48]. A higher hypervolume index indicates a better quality of solution.

The changes of the hypervolume index with generations can provide a qualitative estimation of the optimizing efficiency. For an ideal optimization, the hypervolume index will increase rapidly at the beginning of the optimization, then increase slowly, and eventually remain stable. Therefore, the faster the hypervolume index reaches stability, the better the optimizing efficiency. For convenience and consistency of comparison, a criterion is adopted to judge whether the hypervolume index has reached stability in this study, i.e., if the increment rate of the hypervolume index compared to the former generation is lower than 0.1% for three consecutive generations and there are also no three consecutive increment rates greater than 0.1% in the following generations. For comparability in this study, the worst reference point for calculating the hypervolume index of all experiments was set to (300, 0), which represents the net cost being CNY 3 million and the reduction rate of soil erosion being zero.

Note that both the near-optimal Pareto solutions and hypervolume index represent evaluations from a mathematical perspective, which have less practical meaning than the practicality of the spatial distribution of BMP scenarios for decision-making in integrated watershed management [2]. Therefore, the practicality of the spatial distributions of selected near-optimal Pareto solutions from different BMP configuration units with the corresponding optimal BMP configuration strategy will be qualitatively discussed based on the visual interpretation and local experiences of the study area.

3. Results

3.1. Comparison among Different BMP Configuration Units with the RAND Strategy

Figure 6 shows the near-optimal Pareto solutions of the 100th generation (Figure 6a) and the hypervolume index with generations (Figure 6b) from four types of BMP configuration units applied with the random configuration strategy (i.e., SLPPOS+RAND, CONNFIELD+RAND, EXPLICITHRU+RAND, and HRU+RAND). Effective but different near-optimal Pareto solutions are generated by all combinations in almost the same solution space, i.e., soil erosion reduction rates range from 0.10 to 0.50 with the net cost range from CNY 0.03 to 1.50 million (Figure 6a). The near-optimal Pareto solutions of SLPPOS+RAND and EXPLICITHRU+RAND are nearly overlapped, while HRU+RAND produced the most non-dominated solutions at the nearly entire solution space and CONNFIELD+RAND was dominated by the other three combinations. HRU+RAND achieved the best overall performance considering convergence and diversity, compared to the other three combinations with the RAND strategy (Figure 6a). All four of these combinations obtained approximately the same values of the hypervolume index with generations, which showed a similar change trend, i.e., the hypervolume index increased rapidly for about the first 20 generations (e.g., with increment rates of the hypervolume index greater than 1%) and then increased slowly until stability was reached at the 46th, 41th, 56th, and 68th generations for SLPPOS+RAND, CONNFIELD+RAND, EXPLICITHRU+RAND, and HRU+RAND, respectively (Figure 6b), among which CONNFIELD+RAND showed the best optimizing efficiency.

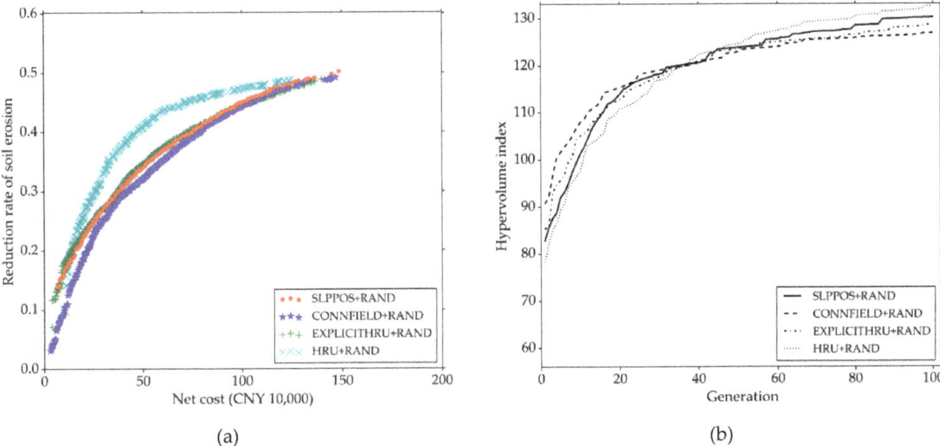

Figure 6. Comparisons among four types of BMP configuration units applied with the random configuration (RAND) strategy: (**a**) near-optimal Pareto solutions of the 100th generation and, (**b**) the hypervolume index with generations.

As the BMP configuration strategy remained the same, the characteristics of BMP configuration units are the main causes of the differences among the optimization results. Compared to the other three types of configuration units, HRUs in the study area obtained the largest count of units and the most detailed spatial delineation. This means that under the RAND strategy, a small number of HRUs configured with BMPs (i.e., at a low net cost) at some critical erosion zone may result in a relatively higher soil erosion reduction rate (i.e., better cost-effectiveness). The generalized hydrologically connected fields with upstream–downstream relationships at the watershed scale usually have large areas and may occupy most of the upslope positions of subbasins, where the critical erosion zone is often located. Therefore, CONNFIELD+RAND performed less effectively at low net costs than the other three combinations and obtained relatively stable high soil erosion reduction rates with relative greater net costs (Figure 6a). Since, in the study area, both landuse and soil types have a similar spatial pattern to the topography (Section 2.2), and spatially explicit HRUs were overlaid by landuse, soil types, and subbasin boundaries, EXPLICITHRU can represent the spatial distribution of topographic characteristics to some degree (Figure 5b). Considering that the slope position units were totally delineated according to topographic characteristics, the similarity between these two spatial configuration units maybe the reason for the quite similar results between EXPLICITHRU+RAND and SLPPOS+RAND. With consistent crossover and mutation operations during optimization, the hypervolume index trends with generations are very similar among four types of BMP configuration units applied with the RAND strategy (Figure 5b). It might be inferred that the optimizing efficiency under the RAND strategy is negatively correlated with the count of BMP configuration units.

3.2. Comparison among Different BMP Configuration Units with the SUIT Strategy

Figure 7 shows a comparison among four types of BMP configuration units applied with the SUIT strategy (i.e., SLPPOS+SUIT, CONNFIELD+SUIT, EXPLICITHRU+SUIT, and HRU+SUIT). The solution spaces of the spatial optimization of EXPLICITHRU+SUIT and HRU+SUIT were inclined to reach higher net costs, while SLPPOS+SUIT and CONNFIELD+SUIT comparatively concentrated on the solution spaces with low net costs (Figure 7a). HRU+SUIT and SLPPOS+SUIT obtained the most non-dominated solutions when the corresponding net costs were greater or less than about CNY 1.10 million, respectively. For HRU+SUIT and EXPLICITHRU+SUIT, the soil erosion reduction rate reached relative stability at about 0.52 when the corresponding net costs were greater than CNY 1.60 million (Figure 7a). The solutions of SLPPOS+SUIT obtained very similar hypervolume index values to those from CONNFIELD+SUIT in the first stage of optimization (about the first 35 generations) and then a higher hypervolume index during the following generations (Figure 7b). The EXPLICITHRU+SUIT and HRU+SUIT produced worse diversity than SLPPOS+SUIT and CONNFIELD+SUIT (Figure 7a). This phenomenon can also be observed from the lower hypervolume index from EXPLICITHRU+SUIT and HRU+SUIT (Figure 7b). The SLPPOS+SUIT, CONNFIELD+SUIT, EXPLICITHRU+SUIT, and HRU+SUIT reached stability at the 46th, 36th, 45th, and 60th generations, respectively (Figure 7b), among which the CONNFIELD+SUIT showed the best optimizing efficiency.

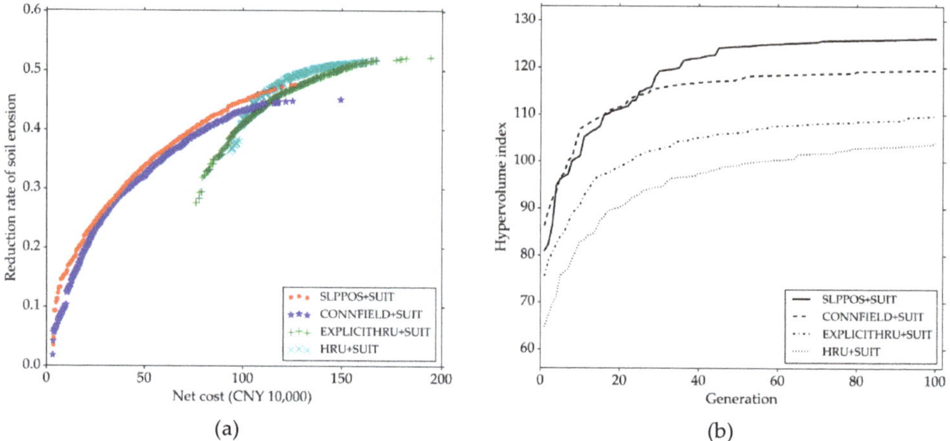

Figure 7. Comparisons among four types of BMP configuration units applied with the suitable landuse types/slope positions of individual BMPs (SUIT) strategy: (**a**) near-optimal Pareto solutions of the 100th generation, and (**b**) the hypervolume index with generations.

Since the HRUs and spatially explicit HRUs delineated in this study have a single landuse type for each spatial configuration unit, the non-effective configurations caused by BMP configuration units with multiple landuse types (e.g., hydrologically connected fields and slope position units) (Section 2.5) do not exist. Therefore, under the same parameter-settings of the optimization algorithm, EXPLICITHRU+SUIT and HRU+SUIT inherently generate more effective BMP configurations in locations within spatial configuration units than SLPPOS+SUIT and CONNFIELD+SUIT and thus result in a solution space with relatively high net costs.

With the best near-optimal Pareto solutions, the highest hypervolume index, and the satisfied optimizing efficiency, SLPPOS+SUIT obtained the best overall performance compared to the other three combinations with the SUIT strategy. This may be attributed to the knowledge used by the SUIT strategy for slope position units that considers not only suitable landuse types but also suitable slope positions of individual BMPs.

3.3. Comparison between Feasible BMP Configuration Units with the UPDOWN Strategy

Figure 8 presents a comparison between two types of BMP configuration units applied with the UPDOWN strategy (i.e., SLPPOS+UPDOWN and CONNFIELD+UPDOWN). CONNFIELD+UPDOWN and SLPPOS+UPDOWN obtained most of their non-dominated solutions when the corresponding net costs were greater or less than about CNY 0.25 million, respectively (Figure 8a). SLPPOS+UPDOWN had a narrower solution space than CONNFIELD+UPDOWN and hence had worse diversity and a lower hypervolume index. CONNFIELD+UPDOWN showed a more steady hypervolume index trend with generations and better optimizing efficiency since it reached stability at the 39th generation which is lower than the 55th generation for SLPPOS+UPDOWN (Figure 8b). SLPPOS+UPDOWN reached slightly better hypervolume index stability after about the 70th generation, i.e., a slightly better convergence than CONNFIELD+UPDOWN.

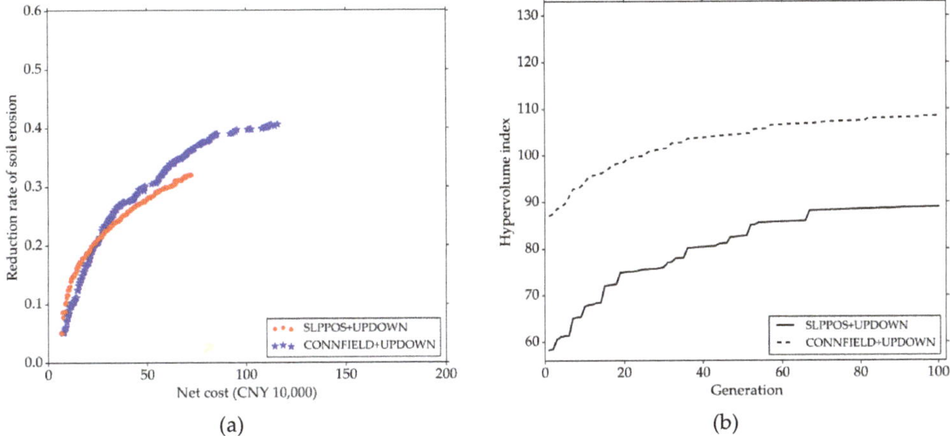

Figure 8. Comparisons between two types of BMP configuration units applied with the UPDOWN strategy: (**a**) near-optimal Pareto solutions of the 100th generation, and (**b**) the hypervolume index with generations.

Although the same strategy was applied, the upstream–downstream relationships of slope position units were built at the hillslope scale instead of the watershed scale of hydrologically connected fields. This means that the number of slope position units within one hillslope (i.e., three in this study) is the maximum number of genes allowed to be exchanged during crossover operation in NSGA-II (Section 2.6). This puts the SLPPOS+UPDOWN result in a slightly better convergence and worse diversity than CONNFIELD+UPDOWN. The non-dominated solutions with low net costs of SLPPOS+UPDOWN indicated the better cost-effectiveness of slope position units under a tight budget, while its worse cost-effectiveness with higher net costs than CONNFIELD+UPDOWN may imply that the UPDOWN strategy initially developed for hydrologically connected fields [5] is not the most effective strategy for slope position units.

3.4. Comparison among Different BMP Configuration Units with the Corresponding Optimal Configuration Strategies

As shown in Figure 9a, HRU+SUIT and SLPPOS+HILLSLP generated the most effective non-dominated solutions when the corresponding net costs were greater and less than about CNY 1.0 million, respectively. According to the hypervolume index (Figure 9b), SLPPOS+HILLSLP reached the stable hypervolume index after the 37th generation, as well as the largest hypervolume index value, compared to the other three combinations. Thus, SLPPOS+HILLSLP showed the best optimizing efficiency. In summary, SLPPOS+HILLSLP showed the best overall performance (Figure 9), followed by EXPLICITHRU+SUIT and CONNFIELD+UPDOWN. Although HRU+SUIT had non-dominant solutions mostly at high net costs, it was still considered to have the worst overall performance because it produced the worst diversity, lowest hypervolume index, and slowest optimizing efficiency (i.e., reaching stability at the 60th generation; Section 3.2).

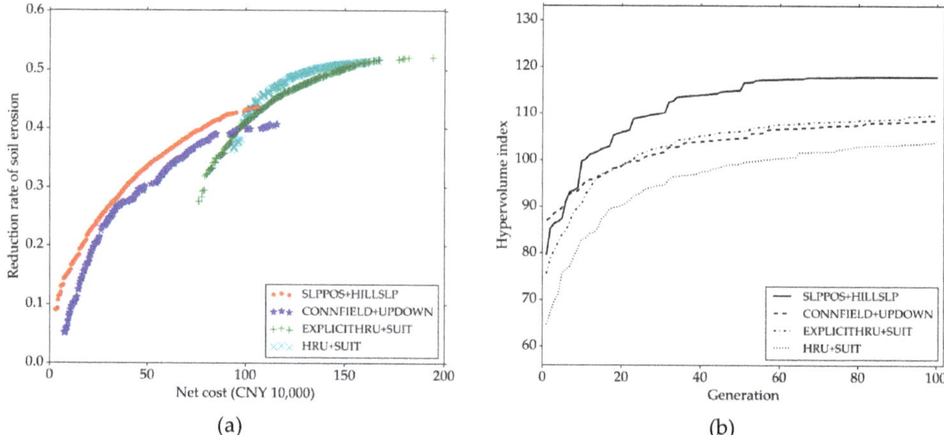

Figure 9. Comparisons among four types of BMP configuration units applied with the corresponding optimal configuration strategies: (**a**) near-optimal Pareto solutions of the 100th generation, and (**b**) the hypervolume index changed with generations.

To compare the practicality of the spatial distribution of optimized BMP scenarios, four BMP scenarios (Figure 10) were selected in the overlapped solution space of the four combinations (i.e., at a soil erosion reduction rate of around 0.40 and net cost of CNY 0.70–1.20 million). The most fragmented spatial delineation of HRUs caused the worst practicality from the perspective of actual watershed management (Figure 10a). EXPLICITHRU+SUIT (Figure 10b) generated similar BMPs regions to HRU+SUIT (Figure 10a), but those from EXPLICITHRU+SUIT had a more concentrated and practical distribution of BMPs than those from HRU+SUIT. The BMP scenarios from CONNFIELD+UPDOWN required the highest net cost to achieve the same soil erosion reduction rate among these selected BMP scenarios, since CONNFIELD+UPDOWN selected the most expensive BMP (i.e., Orchard Improvement) for allocation (Figure 10c). With the adoption of the UPDOWN strategy and a smaller configuration area, the BMP scenario from CONNFIELD+UPDOWN (Figure 10c) achieved better practicality than that from EXPLICITHRU+SUIT (Figure 10b). With the SUIT and UPDOWN strategies, the lack of precise spatial relationships between BMPs and spatial positions at the hillslope scale may cause several inappropriate configurations (Figure 10a–c), e.g., the LQFI (Low-quality forest improvement) configured on ridge areas (Table 1), thus reducing the practicality of the BMP scenario from EXPLICITHRU+SUIT and CONNFIELD+UPDOWN. Depending on the HILLSLP strategy used to adopt BMP knowledge derived from the local experience of integrated watershed management, SLPPOS+HILLSLP concisely and precisely configured CM (Closing measures) and ABHMP (Arbor–bush–herb mixed plantation) on slope position units at the hillslope scale, e.g., CM-ABHMP on the ridge and backslope sequence, and also configured the BMP with the best overall effectiveness grade (i.e., ABHMP according to Table 3) on ridge and backslope. Thus, SLPPOS+HILLSLP obtained the best practicality with the lowest net cost among these selected BMP scenarios, followed by CONNFIELD+UPDOWN and EXPLICITHRU+SUIT.

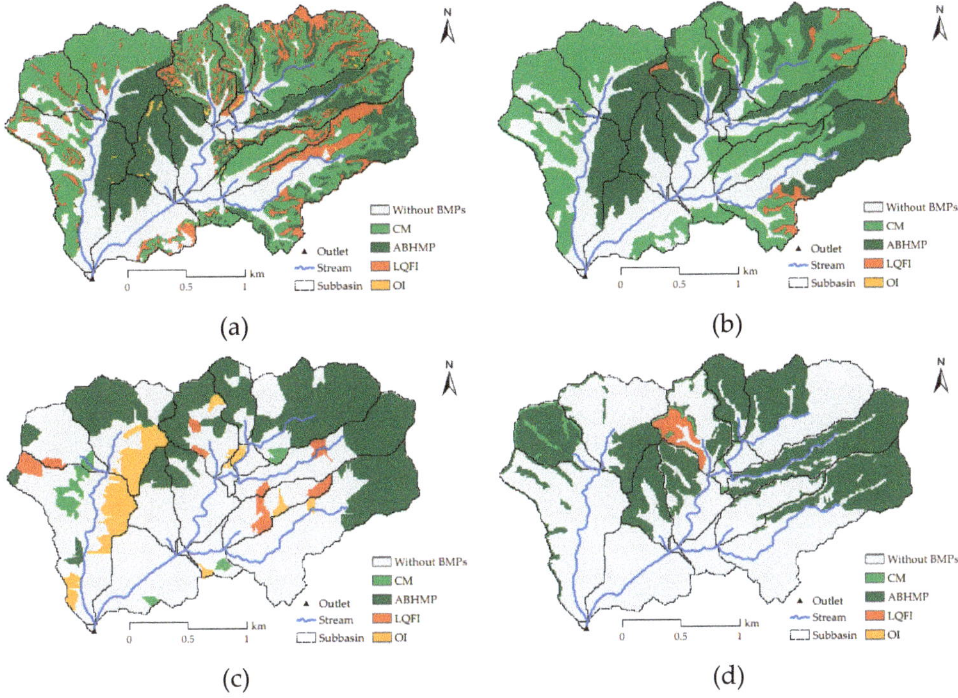

Figure 10. Comparison of the spatial distribution of selected BMP scenarios of the 100th generation from four types of BMP configuration units applied with the corresponding optimal strategies: (**a**) one HRU+SUIT scenario with a soil erosion reduction rate of 0.41 and a net cost of CNY 0.98 million; (**b**) one EXPLICIT+SUIT scenario with a soil erosion reduction rate of 0.40 and a net cost of CNY 0.97 million; (**c**) one CONNFIELD+UPDOWN scenario with a soil erosion reduction rate of 0.40 and a net cost of CNY 1.08 million; and (**d**) one SLPPOS+HILLSLP scenario with a soil erosion reduction rate of 0.40 and a net cost of CNY 0.78 million.

4. Discussion

From a mathematical viewpoint, all feasible combinations of BMP configuration units and BMP configuration strategies generated effective near-optimal Pareto solutions (Figures 6–9). Different delineations of BMP configuration units have different characteristics such as the number of units, the spatial distribution characteristics, and the spatial relationships with homogeneous functional units from the perspective of physical geography. These differences affect the characteristics of the generated BMP scenarios and the search spaces of spatial optimization, thus resulting in the different cost-effectiveness of near-optimal Pareto solutions and optimizing efficiency when applied with the same BMP configuration strategy (Sections 3.1–3.3).

BMP scenarios generated based on different kinds of BMP configuration knowledge had significant differences in the cost-effectiveness of the near-optimal Pareto solutions, optimizing efficiency, and the spatial distribution of the BMP scenarios. Using the strategies that adopt BMP configuration knowledge (i.e., the SUIT, UPDOWN, and HILLSLP strategies), those non-effective configurations generated by the random configuration strategy (Section 2.5) can be avoided for all BMP configuration units. Thus, with the BMP configuration constraint, better convergence, and worse diversity of near-optimal Pareto solutions were obtained, as well as overall lower values of hypervolume index and better optimizing efficiency (e.g., comparing Figure 7, Figure 8, and Figure 9 with Figure 6). Such phenomena could also be observed when the adopted knowledge extended from simple knowledge to knowledge on the spatial relationships between BMPs and spatial positions, i.e., from the SUIT strategy to the UPDOWN strategy or the HILLSLP strategy.

Besides the narrowed solution space of spatial optimization and improved optimizing efficiency, the adoption of domain (or expert) knowledge considering spatial relationships between BMPs and spatial positions, e.g., the UPDOWN and HILLSLP strategies in this study, can also facilitate the optimized BMP solutions with geographical and practical meanings (Section 3.4). The formal representation of such expert knowledge is often introduced according to the characteristics of specific BMP configuration units, which means that one BMP configuration strategy developed associated with one BMP configuration unit may not be effectively applied to another BMP configuration unit, e.g., the ineffective situation in which the UPDOWN strategy was applied to slope position units (Figures 8 and 9).

The random strategy and simple knowledge-based strategy (e.g., the SUIT strategy) may obtain more effective near-optimal BMPs solutions at higher net costs than the strategies with knowledge on the spatial relationships between BMPs and spatial positions (e.g., the UPDOWN and HILLSLP strategies) (Figure 9a). However, the latter type of strategies can effectively represent the local experience of integrated watershed management with better practicality [2] and hence are more valuable for real applications, especially when the expert knowledge on the spatial relationships between BMPs and slope positions along a hillslope can be considered (Figure 10d).

5. Conclusions

This paper presents a comparison among the four main types of BMP configuration units (i.e., HRUs, spatially explicit HRUs, hydrologically connected fields, and slope position units) for watershed BMP scenarios optimization based on the same one distributed watershed model. Four BMP configuration strategies, adopting different kinds of BMP knowledge of the study area, were considered, i.e., the random configuration strategy, the strategy with knowledge on suitable landuse types/slope positions of individual BMPs, the strategy based on expert knowledge of upstream–downstream relationships [5], and the strategy with expert knowledge on the spatial relationships between BMPs and slope positions along the hillslope [2]. A total of 11 experiments of feasible combinations were conducted with the same optimization algorithm (i.e., NSGA-II) and then compared.

The comparison showed that different BMP configuration units applied with different configuration strategies had significant differences in near-optimal Pareto solutions, optimizing efficiency, and spatial distribution of BMP scenarios. Generally, the more the expert (or domain) knowledge was considered, the better the comprehensive cost-effectiveness and practicality of the optimized BMP scenarios, and the better the optimizing efficiency. Therefore, BMP configuration units that support the adoption of expert knowledge on the spatial relationships between BMPs and spatial locations [2,5] (i.e., hydrologically connected fields, and slope position units) are considered to be the most valuable spatial configuration units for watershed BMP scenarios optimization and integrated watershed management. Overall, using the slope position units as BMP configuration units with the HILLSLP strategy, the best comprehensive results of BMP scenarios optimization were obtained.

This study provided a useful reference for the spatial optimization of watershed BMP scenarios with multiple spatially explicit BMPs. For those spatially explicit BMPs, more BMP configuration knowledge derived from local management experiences should be summarized and adopted together with the feasible BMP configuration units for such knowledge (e.g., slope position units) during watershed BMP scenarios optimization.

Author Contributions: Conceptualization, L.-J.Z., C.-Z.Q., and A.-X.Z.; Methodology, L.-J.Z. and C.-Z.Q.; Software, L.-J.Z., J.L., and H.W.; Investigation, L.-J.Z.; Resources, J.L., C.-Z.Q., and A.-X.Z.; Writing—Original Draft Preparation, L.-J.Z. and C.-Z.Q.; Funding Acquisition, C.-Z.Q., and A.-X.Z.

Funding: The work reported was financially supported by the National Natural Science Foundation of China (No. 41871362 and 41431177), the National Key Technology R&D Program (No. 2013BAC08B03-4), the Chinese Academy of Sciences (No. XDA23100503), the Innovation Project of LREIS (No. O88RA20CYA), National Basic

Research Program of China (Project No.: 2015CB954102), PAPD, and the Outstanding Innovation Team in Colleges and Universities in Jiangsu Province. Supports to A-Xing Zhu through the Vilas Associate Award, the Hammel Faculty Fellow Award, and the Manasse Chair Professorship from the University of Wisconsin-Madison are greatly appreciated.

Acknowledgments: The authors thank Zhi-Biao Chen for kindly permitting us to use his photos of the four types of BMPs considered in this study.

Conflicts of Interest: The authors declare no conflict of interest.

References

1. Kalcic, M.M.; Frankenberger, J.; Chaubey, I. Spatial Optimization of Six Conservation Practices Using Swat in Tile-Drained Agricultural Watersheds. *J. Am. Water Resour. Assoc. JAWRA* **2015**, *51*, 956–972. [CrossRef]

2. Qin, C.-Z.; Gao, H.-R.; Zhu, L.-J.; Zhu, A.-X.; Liu, J.-Z.; Wu, H. Spatial optimization of watershed best management practices based on slope position units. *J. Soil Water Conserv.* **2018**, *73*, 504–517. [CrossRef]

3. Sahu, M.; Gu, R.R. Modeling the effects of riparian buffer zone and contour strips on stream water quality. *Ecol. Eng.* **2009**, *35*, 1167–1177. [CrossRef]

4. Veith, T.L.; Wolfe, M.L.; Heatwole, C.D. Optimization Procedure for Cost Effective Bmp Placement at a Watershed Scale. *J. Am. Water Resour. Assoc. JAWRA* **2003**, *39*, 1331–1343. [CrossRef]

5. Wu, H.; Zhu, A.-X.; Liu, J.; Liu, Y.; Jiang, J. Best Management Practices Optimization at Watershed Scale: Incorporating Spatial Topology among Fields. *Water Resour. Manag.* **2018**, *32*, 155–177. [CrossRef]

6. Wang, X.; Amonett, C.; Williams, J.R.; Wilcox, B.P.; Fox, W.E.; Tu, M.-C. Rangeland watershed study using the Agricultural Policy/Environmental eXtender. *J. Soil Water Conserv.* **2014**, *69*, 197–212. [CrossRef]

7. Duriancik, L.F.; Bucks, D.; Dobrowolski, J.P.; Drewes, T.; Eckles, S.D.; Jolley, L.; Kellogg, R.L.; Lund, D.; Makuch, J.R.; O'Neill, M.P. The first five years of the Conservation Effects Assessment Project. *J. Soil Water Conserv.* **2008**, *63*, 185A–197A. [CrossRef]

8. Turpin, N.; Bontems, P.; Rotillon, G.; Bärlund, I.; Kaljonen, M.; Tattari, S.; Feichtinger, F.; Strauss, P.; Haverkamp, R.; Garnier, M.; et al. AgriBMPWater: Systems approach to environmentally acceptable farming. *Environ. Model. Softw.* **2005**, *20*, 187–196. [CrossRef]

9. Xie, H.; Chen, L.; Shen, Z.; Xie, H.; Chen, L.; Shen, Z. Assessment of Agricultural Best Management Practices Using Models: Current Issues and Future Perspectives. *Water* **2015**, *7*, 1088–1108. [CrossRef]

10. Arabi, M.; Govindaraju, R.S.; Hantush, M.M. Cost-effective allocation of watershed management practices using a genetic algorithm. *Water Resour. Res.* **2006**, *42*, W10429. [CrossRef]

11. Gitau, M.W.; Veith, T.L.; Gburek, W.J. Farm-level optimization of BMP placement for cost-effective pollution reduction. *Trans. ASAE* **2004**, *47*, 1923–1931. [CrossRef]

12. Maringanti, C.; Chaubey, I.; Arabi, M.; Engel, B. Application of a multi-objective optimization method to provide least cost alternatives for NPS pollution control. *Environ. Manag.* **2011**, *48*, 448–461. [CrossRef] [PubMed]

13. Srivastava, P.; Hamlett, J.M.; Robillard, P.D.; Day, R.L. Watershed optimization of best management practices using AnnAGNPS and a genetic algorithm. *Water Resour. Res.* **2002**, *38*. [CrossRef]

14. Chang, C.L.; Chiueh, P.T.; Lo, S.L. Effect of spatial variability of storm on the optimal placement of best management practices (BMPs). *Environ. Monit. Assess.* **2007**, *135*, 383–389. [CrossRef] [PubMed]

15. Chichakly, K.J.; Bowden, W.B.; Eppstein, M.J. Minimization of cost, sediment load, and sensitivity to climate change in a watershed management application. *Environ. Model. Softw.* **2013**, *50*, 158–168. [CrossRef]

16. Qiu, J.; Shen, Z.; Huang, M.; Zhang, X. Exploring effective best management practices in the Miyun reservoir watershed, China. *Ecol. Eng.* **2018**, *123*, 30–42. [CrossRef]

17. Yang, G.X.; Best, E.P.H. Spatial optimization of watershed management practices for nitrogen load reduction using a modeling-optimization framework. *J. Environ. Manag.* **2015**, *161*, 252–260. [CrossRef]

18. Panagopoulos, Y.; Makropoulos, C.; Mimikou, M. Decision support for diffuse pollution management. *Environ. Model. Softw.* **2012**, *30*, 57–70. [CrossRef]

19. Rodriguez, H.G.; Popp, J.; Maringanti, C.; Chaubey, I. Selection and placement of best management practices used to reduce water quality degradation in Lincoln Lake watershed. *Water Resour. Res.* **2011**, *47*. [CrossRef]

20. Band, L.E. Spatial hydrography and landforms. In *GIS: Management Issues and Applications*; Longley, P., Goodchild, M., Maguire, D., Rhind, D., Eds.; John Wiley & Sons: Hoboken, NI, USA, 1999; pp. 527–542.

21. Jiang, P.; Anderson, S.H.; Kitchen, N.R.; Sadler, E.J.; Sudduth, K.A. Landscape and conservation management effects on hydraulic properties of a claypan-soil toposequence. *Soil Sci. Soc. Am. J.* **2007**, *71*, 803–811. [CrossRef]

22. Mudgal, A.; Baffaut, C.; Anderson, S.H.; Sadler, E.J.; Thompson, A.L. APEX model assessment of variable landscapes on runoff and dissolved herbicides. *Trans. ASABE* **2010**, *53*, 1047–1058. [CrossRef]

23. Arnold, J.G.; Srinivasan, R.; Muttiah, R.S.; Williams, J.R. Large area hydrologic modeling and assessment part I: Model development. *J. Am. Water Resour. Assoc. JAWRA* **1998**, *34*, 73–89. [CrossRef]

24. Arnold, J.G.; Allen, P.M.; Volk, M.; Williams, J.R.; Bosch, D.D. Assessment of Different Representations of Spatial Variability on SWAT Model Performance. *Trans. ASABE* **2010**, *53*, 1433–1443. [CrossRef]

25. Kalcic, M.M.; Chaubey, I.; Frankenberger, J. Defining Soil and Water Assessment Tool (SWAT) hydrologic response units (HRUs) by field boundaries. *Int. J. Agric. Biol. Eng.* **2015**, *8*, 69–80.

26. Teshager, A.D.; Gassman, P.W.; Secchi, S.; Schoof, J.T.; Misgna, G. Modeling Agricultural Watersheds with the Soil and Water Assessment Tool (SWAT): Calibration and Validation with a Novel Procedure for Spatially Explicit HRUs. *Environ. Manag.* **2016**, *57*, 894–911. [CrossRef]

27. Gaddis, E.J.B.; Voinov, A.; Seppelt, R.; Rizzo, D.M. Spatial Optimization of Best Management Practices to Attain Water Quality Targets. *Water Resour. Manag.* **2014**, *28*, 1485–1499. [CrossRef]

28. Limbrunner, J.F.; Vogel, R.M.; Chapra, S.C.; Kirshen, P.H. Optimal Location of Sediment-Trapping Best Management Practices for Nonpoint Source Load Management. *J. Water Resour. Plan. Manag.* **2013**, *139*, 478–485. [CrossRef]

29. Perez-Pedini, C.; Limbrunner, J.F.; Vogel, R.M. Optimal Location of Infiltration-Based Best Management Practices for Storm Water Management. *J. Water Resour. Plan. Manag.* **2005**, *131*, 441–448. [CrossRef]

30. Qin, C.-Z.; Zhu, A.-X.; Shi, X.; Li, B.-L.; Pei, T.; Zhou, C.-H. Quantification of spatial gradation of slope positions. *Geomorphology* **2009**, *110*, 152–161. [CrossRef]

31. Zhu, L.-J.; Zhu, A.-X.; Qin, C.-Z.; Liu, J.-Z. Automatic approach to deriving fuzzy slope positions. *Geomorphology* **2018**, *304*, 173–183. [CrossRef]

32. Bieger, K.; Arnold, J.G.; Rathjens, H.; White, M.J.; Bosch, D.D.; Allen, P.M.; Volk, M.; Srinivasan, R. Introduction to SWAT+, A Completely Restructured Version of the Soil and Water Assessment Tool. *JAWRA J. Am. Water Resour. Assoc.* **2017**, *53*, 115–130. [CrossRef]

33. Zhu, L.-J.; Liu, J.; Qin, C.-Z.; Zhu, A.-X. A modular and parallelized watershed modeling framework. *Environ. Model. Softw.*. under review.

34. Deb, K.; Pratap, A.; Agarwal, S.; Meyarivan, T. A fast and elitist multiobjective genetic algorithm: NSGA-II. *IEEE Trans. Evol. Comput.* **2002**, *6*, 182–197. [CrossRef]

35. Chen, Z.-B.; Chen, Z.-Q.; Yue, H. *Comprehensive Research on Soil and Water Conservation in Granite Red Soil Region: A Case Study of Zhuxi Watershed, Changting County, Fujian Province*; Science Press: Beijing, China, 2013. (In Chinese)

36. Shi, X.; Yang, G.; Yu, D.; Xu, S.; Warner, E.D.; Petersen, G.W.; Sun, W.; Zhao, Y.; Easterling, W.E.; Wang, H. A WebGIS system for relating genetic soil classification of China to soil taxonomy. *Comput. Geosci.* **2010**, *36*, 768–775. [CrossRef]

37. Arnold, J.G.; Kiniry, J.R.; Srinivasan, R.; Williams, J.R.; Haney, E.B.; Neitsch, S.L. *Soil and Water Assessment Tool 2012 Input/Output Documentation*; Texas Water Resources Institute: College Station, TX, USA, 2012.

38. Chen, S.; Zha, X. Evaluation of soil erosion vulnerability in the Zhuxi watershed, Fujian Province, China. *Nat. Hazards* **2016**, *82*, 1589–1607. [CrossRef]

39. Dile, Y.T.; Daggupati, P.; George, C.; Srinivasan, R.; Arnold, J. Introducing a new open source GIS user interface for the SWAT model. *Environ. Model. Softw.* **2016**, *85*, 129–138. [CrossRef]

40. Chen, Z.-Q.; Chen, Z.-B.; Bai, L.; Zeng, Y. A catastrophe model to assess soil restoration under ecological restoration in the red soil hilly region of China. *Pedosphere* **2017**, *27*, 778–787. [CrossRef]

41. Fujian Soil and Water Conservation Monitoring Station; Fujian Agriculture and Forestry University; Fujian Normal University; Changting Soil and Water Conservation Monitoring Station. *Monitoring Report of Soil and Water Loss in Changting County*; Fuzhou, Fujian, China, 2010. (In Chinese)

42. Liu, Y.; Engel, B.A.; Flanagan, D.C.; Gitau, M.W.; McMillan, S.K.; Chaubey, I.; Singh, S. Modeling framework for representing long-term effectiveness of best management practices in addressing hydrology and water quality problems: Framework development and demonstration using a Bayesian method. *J. Hydrol.* **2018**, *560*, 530–545. [CrossRef]

43. Liu, J.; Zhu, A.-X.; Qin, C.-Z.; Wu, H.; Jiang, J. A two-level parallelization method for distributed hydrological models. *Environ. Model. Softw.* **2016**, *80*, 175–184. [CrossRef]

44. Williams, J.R. Sediment-yield prediction with universal equation using runoff energy factor. In *Present and Prospective Technology for Predicting Sediment Yield and Sources*; U.S. Department of Agriculture: Washington, DC, USA, 1975; Volume ARS-S-40, pp. 244–252.

45. Morris, M.D. Factorial sampling plans for preliminary computational experiments. *Technometrics* **1991**, *33*, 161–174. [CrossRef]

46. Moriasi, D.N.; Arnold, J.G.; Van Liew, M.W.; Bingner, R.L.; Harmel, R.D.; Veith, T.L. Model evaluation guidelines for systematic quantification of accuracy in watershed simulations. *Trans. ASABE* **2007**, *50*, 885–900. [CrossRef]

47. Zitzler, E.; Thiele, L. Multiobjective evolutionary algorithms: A comparative case study and the strength Pareto approach. *IEEE Trans. Evol. Comput.* **1999**, *3*, 257–271. [CrossRef]

48. Zitzler, E.; Thiele, L.; Laumanns, M.; Fonseca, C.M.; da Fonseca, V.G. Performance assessment of multiobjective optimizers: An analysis and review. *IEEE Trans. Evol. Comput.* **2003**, *7*, 117–132. [CrossRef]

Article

Evaluating the Effectiveness of Spatially Reconfiguring Erosion Hot Spots to Reduce Stream Sediment Load in an Upland Agricultural Catchment of South Korea

Kwanghun Choi [1,2,*], **Ganga Ram Maharjan** [3,4] **and Björn Reineking** [1,5]

1. Biogeographical Modelling, Bayreuth Center of Ecology and Environmental Research (BayCEER), University of Bayreuth, Universitätsstraße 30, 95440 Bayreuth, Germany; bjoern.reineking@irstea.fr
2. Department of Forestry, Environment, and Systems, Kookmin University, Seoul 02707, Korea
3. Department of Soil Physics, University of Bayreuth, Universitätsstraße 30, 95440 Bayreuth, Germany; mhjgangaram@gmail.com
4. YARA-Crop Nutrition Research and Development, Hanninghof 35, 48249 Dülmen, Germany
5. Univ. Grenoble Alpes, Irstea, LESSEM, 38000 Grenoble, France
*   Correspondence: kchoi@kookmin.ac.kr or kwanghun.choi@yahoo.com

Received: 30 March 2019 ; Accepted: 30 April 2019; Published: 7 May 2019

Abstract: Upland agricultural expansion and intensification cause soil erosion, which has a negative impact on the environment and socioeconomic factors by degrading the quality of both nutrient-rich surface soil and water. The Haean catchment is a well-known upland agricultural area in South Korea, which generates a large amount of sediment from its cropland. The transportation of nutrient-rich sediment to the stream adversely affects the water quality of the Han River watershed, which supports over twenty million people. In this paper, we suggest a spatially explicit mitigation method to reduce the amount of sediment yield to the stream of the catchment by converting soil erosion hot spots into forest. To evaluate the effectiveness of this reconfiguration, we estimated the sediment redistribution rate and assessed the soil erosion risk in the Haean catchment using the daily based Morgan–Morgan–Finney (DMMF) model. We found that dry crop fields located in the steep hill-slope suffer from severe soil erosion, and the rice paddy, orchard, and urban area, which are located in a comparatively lower and flatter area, suffer less from erosion. Although located in the steep hill-slope, the forest exhibits high sediment trapping capabilities in this model. When the erosion-prone crop lands were managed by sequentially reconfiguring their land use and land cover (LULC) to the forest from the area with the most severe erosion to the area with the least severe erosion, the result showed a strong reduction in sediment yield flowing to the stream. A change of 3% of the catchment's crop lands of the catchment into forest reduced the sediment yield entering into the stream by approximately 10% and a change of 10% of crop lands potentially resulted in a sediment yield reduction by approximately 50%. According to these results, identifying erosion hot spots and managing them by reconfiguring their LULC is effective in reducing terrestrial sediment yield entering into the stream.

Keywords: DMMF; landscape configuration; landscape ecology; hydrology

---

1. Introduction

Agriculture expansion and intensification often lead to severe soil erosion in the course of altering naturally dominated surface configurations [1–3]. The problem is prominent in upland agriculture areas under monsoonal climate because of the disturbed erosion-prone hill-slopes receiving intermittent concentrated heavy rainfall [4,5]. A large amount of surface runoff from heavy rainfall washes out nutrient-rich surface soil from deforested upland agriculture areas and degrades the soil quality of the agricultural area [6]. Eroded nutrient-rich soil particles cause not only soil quality degradation of the agricultural area but also on- and off-site water deterioration when these particles enter the stream of a catchment [7–9].

The Han River watershed in South Korea experiences extreme downpours that cause severe soil erosion and subsequent water deterioration every summer monsoon season [3,10,11]. These problems are worsening, as upland agricultural areas expand and the intensity of monsoonal rainfall increase due to ongoing climate change [12,13]. The Han River is the primary freshwater source for the Seoul Metropolitan area where over 25 million inhabitants (ca. 50% of the South Korean population) reside. Therefore, soil erosion control in this region is highly relevant to provide clean and usable freshwater resources to the residents [14,15]. With increasing demand for food crops, intensive upland agriculture is expanding in the mountainous upstream regions of the Han River watershed where few agricultural activities had been performed previously [2]. The Haean catchment is one of the largest contributors to sediment in the watershed, where abrupt land use and land cover (LULC) changes have taken place on forested hill-slope areas [11,16,17]. The LULC changes on the erosion-prone hill-slopes of this catchment generate a massive amount of sediment flowing into the river system and eventually deteriorate the water quality of the Han River [3]. Various studies have been conducted in this catchment to understand the sediment redistribution patterns and determine optimal measures to mitigate this problem. Field-level studies have focused on the effect of surface configurations of the dry croplands and their field margins on sediment yields. Arnhold et al. [11] and Ruidisch et al. [16] investigated the effect of plastic mulch applied to dry croplands on surface runoff and sediment yield. Ali and Reineking [5] showed the effectiveness of natural field margin (i.e., vegetated filter strip next to the dry cropland) for preventing off-site sediment yield. They reported that the natural field margin captured sediments more efficiently under the increased rainfall and slope conditions than intensively managed field margins with less dense vegetation cover. Arnhold et al. [17] found that organic farming yielded less sediment than conventional farming because organic farming tends to protect the soil surface by preserving more vegetations that are not cultivated crops.

At the catchment level, the soil and water analysis tool (SWAT) [18] has been widely used to test the effectiveness of various best management practices (BMPs) to reduce the sediment yield under complex terrain and landscape configurations [3,19]. Maharjan et al. [3] showed the effectiveness of catchment-wide cover crop cultivation in the dry croplands to reduce suspended sediment yields entering the stream. Jang et al. [19] projected vegetation filter strip, rice straw mulching, and fertilizer control scenarios to dry croplands of the catchment and found that the application of vegetation filter strips and rice straw mulching was efficient in reducing sediment yields from the catchment. The BMPs suggested in the aforementioned studies are often premised on the compliance of each stakeholder, which is not easily accomplished [20–22]. Different from the BMP approaches relying on stakeholders participation, several studies are paying attention to the importance of the landscape and its spatial configuration, which has a significant impact on ecosystem services and functions, including soil erosion and water quality control [23–25]. Furthermore, these studies showed that ecosystem services and functions often responded non-linearly to the spatial relocation of the agricultural landscape, implying the effectiveness of spatial configuration on enhancing ecosystem services [23,24,26]. Therefore, identifying soil erosion hot spots and assessing the sediment reduction rate by altering the surface configuration of hot spots promise to help establishing cost-effective soil erosion control methods in the catchment.

To consider the spatial context of soil erosion, a spatially explicit and distributed soil erosion model that can simulate the sediment budget of each element, considering the sediment inputs from the upslope areas is needed. Among the various soil erosion models, the daily based Morgan–Morgan–Finney (DMMF) model [15] is one of the most appropriate tools because the model can project soil erosion and deposition explicitly, considering the spatial connectivity, which facilitates the assessment of the impact of the spatial context of landscape on sediment redistribution patterns. Furthermore, the DMMF is suitable for projecting under a monsoon climate, accompanying concentrated rainfall during a short period [15]. Vegetative filter strips (VFSs) are known as an effective tool for reducing sediment yield from the field or catchment because of their cost-effective surface protecting and sediment trapping capabilities [5,19,25,27–29]. We adopt the forest, which is a type of VFS, as an alternative LULC for soil erosion hot spots to reduce the total sediment yield into the stream of the catchment. In this study, we assessed the importance of the spatial conversion of erosion hot spots into forest on soil erosion control using the spatially explicit daily based Morgan–Morgan–Finney (DMMF) soil erosion model. The detailed objectives are to:

1. determine the applicability of the DMMF model for stream discharge and suspended sediment in the Haean catchment,
2. estimate the sediment redistribution pattern and assess the soil erosion risk of the Haean catchment, and
3. evaluate the impact of the spatial reconfiguration of erosion hot spots into forest on soil erosion control.

2. Materials and Methods

2.1. Study Area

The study was conducted in the Haean catchment (Figure 1). The Haean catchment is a bowl-shaped small mountainous erosion basin (64.4 km^2) located in the northeastern part of South Korea (38.277° N, 128.135° E). As an erosion basin, the central area is low and flat, and it becomes higher and steeper toward the boundary. The lowest altitude of the catchment is 339 m, and the highest one is 1321 m [2,3,11,30]. Geologically, the catchment consists primarily of two bedrocks. One is gneiss at the higher elevation near the catchment boundary, and the other is highly weathered granite at the flat central area [2,30]. Differential erosion between the two bedrocks formed the unique bowl-shaped catchment [2]. The major soil type of the catchment is cambisol from weathered granite. The dominant soil texture of the catchment is loamy sand (59.4%) followed by sandy loam (27.5%), and sand (10.5%), which has a high infiltration capacity [3,30].

The climate of the catchment is characterized by cold and dry winter, affected by the continental Siberian high, and hot and humid summer affected by the subtropical North Pacific high [30–32]. The average annual precipitation from 2009 to 2011 is 1599 mm, and almost 70% of the rainfall is concentrated in the three months from June to August [3,11,19,30]. Due to climate change, the period of rain spell, as well as the frequency and intensity of heavy rainfall, has increased in this region [33,34].

The dominant land cover type of the catchment is forest. Forest mainly covers the summit and upper hill-slope areas around the boundary of the catchment, occupying 58% of the entire catchment area. Dry croplands (22%), including bean, cabbage, potato, radish, and ginseng, dominate the lower hill-slope areas adjacent to the forest edge. Rice paddies (8%) and residential areas (3%) (e.g., roads and artificial structures) occupy the flat central area of the catchment. Semi-natural vegetation field (8%), shrublands (1%), and bare surface (5%), including fallow and barren field, cover the remaining areas [35].

The dry croplands have been expanded into the forest that is located in the hill-slope area. Due to the upland agriculture expansion after deforestation, the catchment yields a massive amount of sediment into the stream during the summer monsoon season. The sediment is transported to the Soyang reservoir. This reservoir is the largest reservoir in South Korea as well as the crucial freshwater source for citizens living in the Seoul metropolitan area [3,11,30]. Weather stations and hydrological measurement facilities are installed in the catchment to monitor the climate and stream conditions, and erosion control dams and the reservoir have been constructed to reduce the sediment yield from the catchment [30,36].

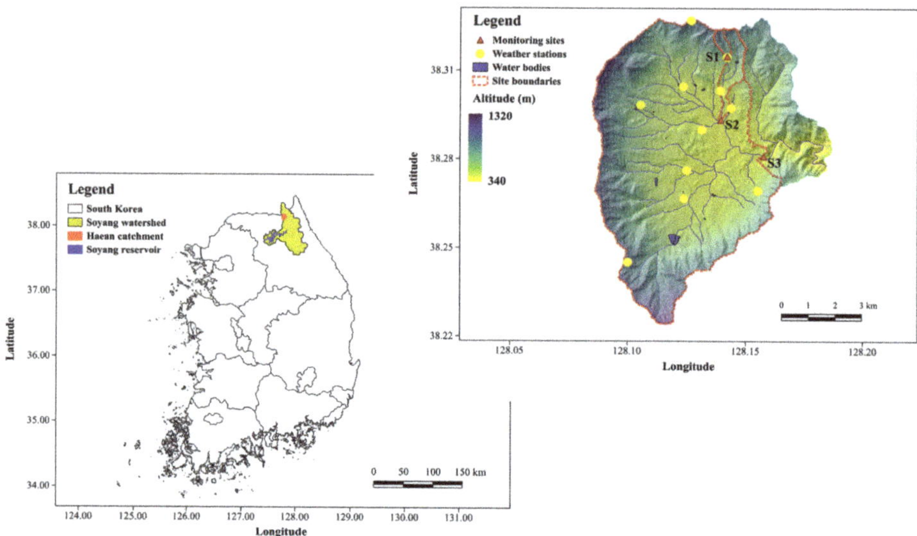

Figure 1. General description of the study area. Locations of the Soyang lake watershed and Haean catchment in South Korea are described in the lower left figure. In the upper right figure, the topography and stream networks of the study area, with the monitoring sites (red triangles) and weather stations (yellow circles) used for the DMMF model are presented.

2.2. Model Description

We used the DMMF model [15] to assess the soil erosion risk and simulate the impact of the spatial reconfiguration of erosion hot spots into forest on sediment yield within the Haean catchment The DMMF model was modified from the widely used Morgan–Morgan–Finney (MMF) soil erosion model [37], which has a simple structure while maintaining physical foundations [15,38–41].

The DMMF model has three significant modifications relative to the MMF model: the adoption of a daily time step, the consideration of the effect of impervious ground cover on soil erosion, and the revision of the equations and sequence of the subprocesses for a better physical representation of physical processes, such as surface runoff and sediment redistribution [15,42]. These modifications enable the model to be more suitable for estimating surface runoff and soil erosion on a complex surface terrain under an intensive seasonal rainfall regime than the previous version.

The DMMF model can estimate the amount of surface and subsurface water input from the upslope area and output to the downslope area after hydrological processes for each element (e.g., each grid cell in a raster map). The model also estimates the sediment budget of each element by calculating the amount of sediments flowing into and out of the element. The hydrological processes of the model are determined by rainfall, evapotranspiration, surface/subsurface water inflows, and initial soil water content (Figure 2). After calculating the water budget for the element, the model calculates sediment budgets, considering the amount of sediment input from the upslope areas, rainfall intensity, topography, soil characteristics, surface configurations, and vegetation structures (Figure 3). The detailed input parameters are presented in Table 1 and detailed structure and equations are described in the Appendix A.

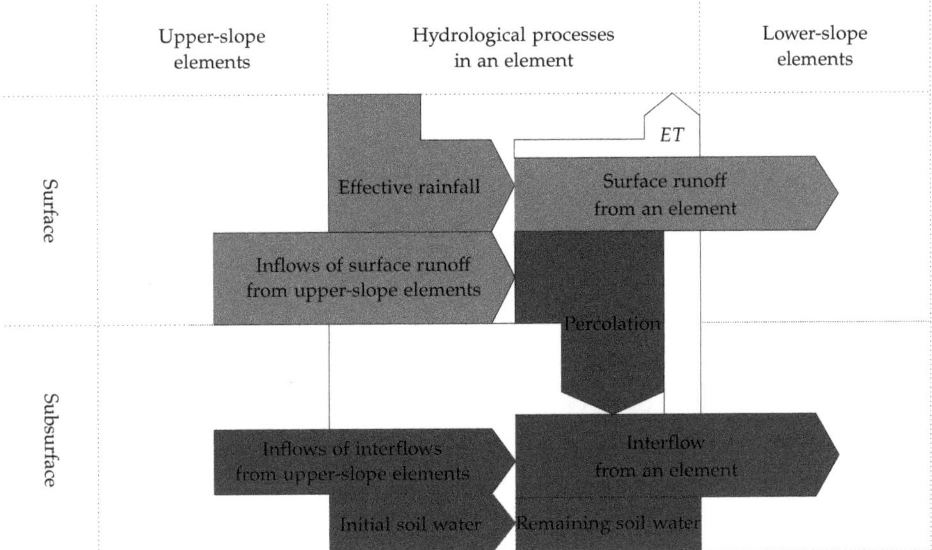

Figure 2. Schematic hydrological phase of the DMMF model (modified from Figure 3 of Choi et al. [15]).

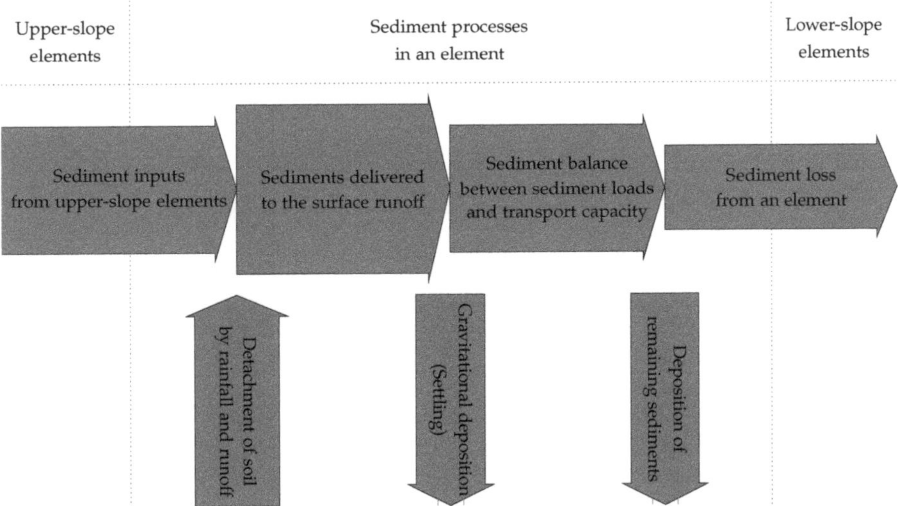

Figure 3. Schematic sediment phase of the DMMF model (modified from Figure 4 of Choi et al. [15]).

In contrast with the SWAT model, which has been frequently applied to this catchment, the DMMF model can estimate the erosion and deposition of an element, considering the interconnectivity with adjacent elements. Therefore, the model can be used to estimate the impact of the spatial reconfiguration of erosion hot spots into forest on sediment yields more explicitly for each element and the entire catchment.

Table 1. Input parameters of the daily based Morgan–Morgan–Finney (DMMF) model (modified from Table 1 of Choi et al. [15])

Type	Parameter	Description	Unit
Topography	S	Slope angle	(rad)
	res	Grid size of a raster map	(m)
Climate	R	Daily rainfall	(mm/day)
	RI	Mean rainfall intensity of a day	(mm/h)
	ET	Daily evapotranspiration	(mm/day)
Soil	P_c	Proportion of clay in the surface soil	(proportion)
	P_z	Proportion of silt in the surface soil	(proportion)
	P_s	Proportion of sand in the surface soil	(proportion)
	SD	Soil depth	(m)
	θ_{init}	Initial soil water content of the entire soil profile	(vol/vol)
	θ_{sat}	Saturated water content of the entire soil profile	(vol/vol)
	θ_{fc}	Soil water content at field capacity of the entire soil profile	(vol/vol)
	K	Saturated soil lateral hydraulic conductivity of the entire soil profile	(mm/day)
	DK_c	Detachability of clay particles by rainfall	(g/J)
	DK_z	Detachability of silt particles by rainfall	(g/J)
	DK_s	Detachability of sand particles by rainfall	(g/J)
	DR_c	Detachability of clay particles by surface runoff	(g/mm)
	DR_z	Detachability of silt particles by surface runoff	(g/mm)
	DR_s	Detachability of sand particles by surface runoff	(g/mm)
LULC	PI	Area proportion of the permanent interception of rainfall	(proportion)
	IMP	Area proportion of the impervious ground cover	(proportion)
	GC	Area proportion of the pervious ground cover of the soil surface	(proportion)
	CC	Area proportion of the canopy cover of the soil surface	(proportion)
	PH	Average height of vegetation or crop cover	(m)
	D	Average diameter of individual plant elements at the surface	(m)
	NV	Number of individual plant elements per unit area	(number/m^2)
	d_a	Typical flow depth of surface runoff	(m)
	n	Manning's roughness coefficient of the soil surface	(s/m$^{1/3}$)

2.3. Model Parameterization

As shown in Table 1, the DMMF model requires the topography, climate, soil, and LULC datasets to project surface runoff and sediment redistribution patterns of the catchment.

Topography data (i.e., the slope angle (S) and grid size of a raster map (res)) were derived from the digital elevation model (DEM) with 30 m resolution. The parameter res is used to calculate the width (w) and length (l) of an element that are equivalent to res and $res/\cos(S)$, respectively [15].

Climate data were obtained from two sources. The daily rainfall (R) and mean rainfall intensity of a day (RI) were obtained from weather stations installed in the catchment, and the evapotranspiration (ET) was obtained from remote sensing data provided by the Moderate Resolution Imaging Spectroradiometer (MODIS) [43]. We estimated R and RI from each weather station and spatially interpolated them using inverse distance weighted (IDW) method, which showed the optimal result on this catchment among four methods such as inverse distance weighted, spline, nearest neighbor, and kriging, according to Shope et al. [30]. For the ET, we resampled the 8-day average MODIS/Terra Evapotranspiration data to fit to the DEM of this catchment.

The soil data set covers the texture, depth, hydraulic properties, and detachabilities. The soil texture (i.e., the proportion of clay (P_c), silt (P_z), and sand (P_s) in the surface soil), soil depth (SD), and soil hydraulic properties (i.e., saturated soil water content (θ_{sat}), soil water content at field capacity (θ_{fc}), and saturated lateral hydraulic conductivity (K) of the entire soil profile) were derived from a 2009 catchment-wide field survey from the TERRECO project (see Table 2 and Figure 4) [30].

Table 2. Typical soil characteristics of each represented soil class of the Haean catchment from a 2009 catchment-wide field survey from TERRECO project.

Classification	SD	P_c	P_z	P_s	θ_{sat} *	θ_{fc} *	K *
Very steep forest	2.55	0.17	0.33	0.50	0.47 (0.41–0.53)	0.21 (0.06–0.31)	1.97 (0.63–4.55)
Forest	4.38	0.22	0.35	0.43	0.45 (0.41–0.54)	0.17 (0.06–0.33)	2.18 (0.63–4.55)
Moderate to steep dry field	2.18	0.08	0.29	0.64	0.36 (0.34–0.39)	0.18 (0.17–0.20)	0.33 (0.18–0.66)
Flat dry field	4.85	0.03	0.15	0.82	0.36 (0.34–0.41)	0.18 (0.08–0.25)	0.49 (0.09–2.25)
Rice paddy	1.60	0.07	0.32	0.62	0.37 (0.36–0.39)	0.16 (0.14–0.18)	0.50 (0.41–0.72)
Sealed ground	2.00	1.00	0.00	0.00	-	-	-

* θ_{sat}, θ_{fc}, and K were estimated with the model ROSETTA Lite v.1.1 [44]. The numbers in parentheses indicate the range of values of soil layers that constitute each represented soil class.

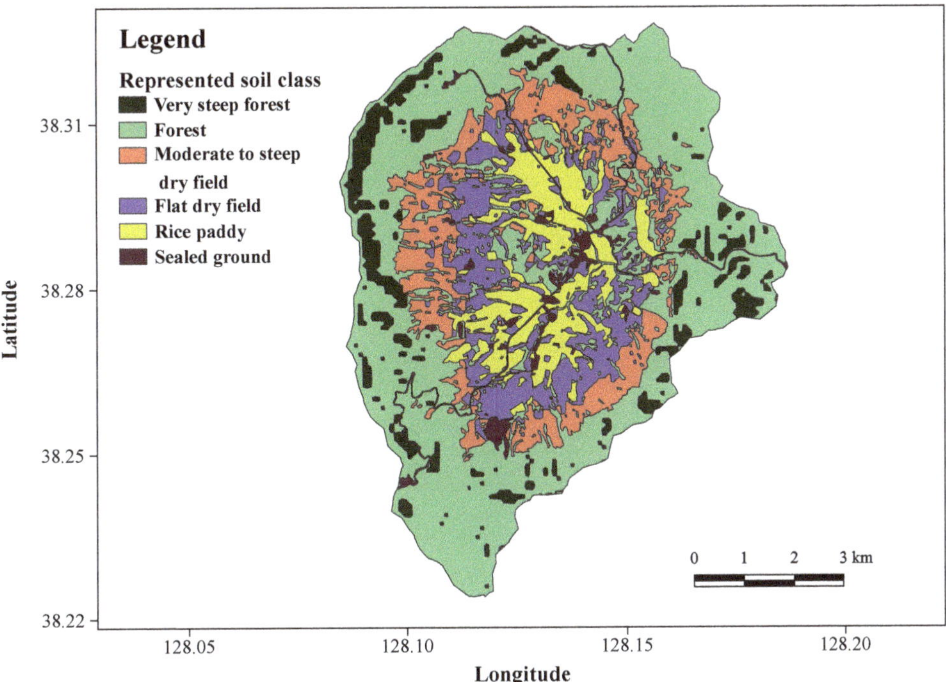

Figure 4. Represented soil class from a 2009 catchment-wide field survey from the TERRECO project.

Reference values for soil detachability from Morgan and Duzant [40] were used as the initial values of soil detachability by rainfall (i.e., for clay (DK_c), silt (DK_z), and sand (DK_s)) and by runoff (i.e., for clay (DR_c), silt (DR_z), and sand (DR_s)). We assumed that the initial soil water content of the entire soil profile (θ_{init}) is equal to the soil water content at field capacity (θ_{fc}) by starting the simulation at three days after the first heavy rainfall of the year, because the excess soil water was usually drained away two or three days after the soil was fully saturated by rainfall.

The LULC types characterize the physical structures of surface and vegetation, which regulate the quantity of surface runoff and runoff velocity. Surface structures incorporate a portion of the impervious cover area (IMP), such as plastic mulching and paved facilities, flow depth of surface runoff (d_a), and Manning's roughness coefficient of the soil surface (n). Vegetation structures contain the permanent interception of rainfall (PI), pervious ground cover (GC), canopy cover (CC), average vegetation height (PH), average diameter of individual plant elements at the surface (D), and number of individual plant elements per unit area (*NV*). LULC parameters were derived based on the LULC map of the Haean catchment in the year 2010 from Seo et al. [35] (see Figure 5).

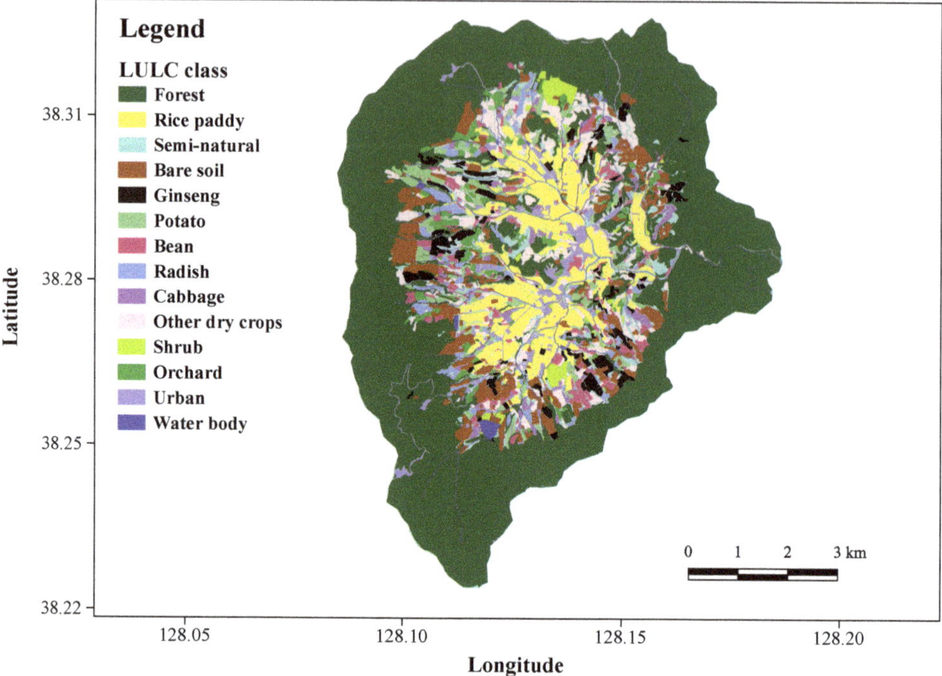

Figure 5. LULC classes and their spatial configurations for the Haean catchment in the year 2010 [35].

We classified the original LULCs into 14 categories (i.e., forest, rice paddy, semi-natural, bare soil, ginseng, potato, bean, radish, cabbage, other dry crops, shrub, orchard, urban, and water bodies). Forest, rice paddy, semi-natural, bare soil, ginseng, potato, bean radish, and cabbage are major LULCs that covered more than 1% of the catchment area. Minor LULCs were aggregated into groups of other dry crops, shrub, orchard, urban, and water bodies according to their physical characteristics. We used field measurement data of *CC*, *PH*, *NV*, *IMP*, d_a, and *n* for major dry crops such as bean, cabbage, potato, and radish, whose data were obtained from the field campaign of the TERRECO project, which was also used in Arnhold et al. [17]. The daily forest *CC* was estimated using the average values of 8-day normalized difference vegetation index (NDVI) for forest in the catchment from MODIS [45,46]. The average NDVI values were converted to canopy cover (*CC*), using the equation suggested by Gutman and Ignatov [47]. LULC parameters for rice and ginseng, and the average diameter of individual plant elements (*D*) for major dry crops were obtained from agricultural technology portal provided by Rural Development Administration of South Korea (RDA) [48]. The average LULC parameters of major dry crops were used for the LULC parameters of other dry crops, while the guide values from Morgan and Duzant [40] were adopted for other LULC parameters. Detailed initial parameter settings are presented in Table 3.

Table 3. The initial parameter settings for each LULC class.

LULC	Leaf-out [a] (Planting)	Leaf-fall [a] (Harvest)	PI [b]	IMP [c]	GC [d]	CC_{max} [e]	PH [f]	D [g]	NV [h]	d_a [i]	n [j]
Forest	112	307	0.20	0.00	1.00	0.95	30.0	2.00	0.60	0.100	0.20
Semi-natural	112	307	0.30	0.00	1.00	0.95	0.50	0.01	500	0.100	0.20
Shrub	112	307	0.20	0.00	0.30	0.95	0.50	0.12	20	0.100	0.20
Rice paddy	136	283	0.30	0.00	1.00 (0.00)	0.80	1.00	0.04	200	0.050	0.10
Potato	120	243	0.12	0.50 (0.00)	0.00 (0.26)	0.71	0.45	0.10	6.00	0.150	0.10
Bean	147	304	0.20	0.50 (0.50)	0.00 (0.58)	0.89	0.70	0.02	6.00	0.150	0.10
Radish	153	235	0.15	0.50 (0.25)	0.00 (0.14)	0.64	0.48	0.06	6.00	0.150	0.10
Cabbage	140	201	0.25	0.50 (0.50)	0.00 (0.31)	0.85	0.55	0.20	3.64	0.150	0.10
Other dry crops	120	304	0.18	0.50 (0.31)	0.00 (0.32)	0.77	0.57	0.10	5.32	0.150	0.10
Orchard	120	303	0.25	0.00	0.40	0.95	4.00	1.50	0.16	0.050	0.10
Ginseng *	123	298	0.20	0.00	0.50	1.00	1.30	0.01	37.5	0.400	0.20
Bare soil	-	-	0.00	0.00	0.00	0.00	0.00	0.00	0.00	0.050	0.01
Urban	-	-	0.00	1.00	0.00	0.00	0.00	0.00	0.00	0.005	0.01

[a] Typical leaf-out and leaf-fall dates of each LULC were presented as day of the year (DOY). For annual crops, the dates represented the typical planting and harvest date of each crop [30]; [b] The reference values from Morgan and Duzant [40] were used for the area proportion of the permanent interception of rainfall (*PI*) for each LULC type; [c] *IMP* for dry fields are different between cultivation and non-cultivation periods. Values in parentheses represent *IMP* for non-cultivation periods; [d] *GC* for dry fields is different before and after harvest. After harvest, crop residues and weeds remained as the ground cover of dry fields, according to dry crop data, from the field campaign of the TERRECO project in 2009. *GC* for rice paddy in cultivation season was set to one reflecting water-filled condition that protected the surface from erosion; [e] Because *CC* values varied with time, we made a list of maximum *CC* (CC_{max}). Semi-natural, shrub, and ginseng utilize fixed reference values from Morgan and Duzant [40]; [f] We used fixed reference *PH* values from Morgan and Duzant [40] for LULCs of other than dry crops. Maximum *PH* values for dry crops were listed from the field measurement data varying with time; [g] We used fixed reference *D* values from Morgan and Duzant [40] for LULCs of other than dry crops. *D* values for dry crops utilized typical crop characteristics from Rural Development Administration of South Korea [48]; [h] We used reference *NV* values from Morgan and Duzant [40] for LULCs of other than dry crops and ginseng. *NV* values which were estimated from the field measurement data and Rural Development Administration of South Korea [48] were used for dry crops and ginseng, respectively; [i] We assumed shallow rill condition for forest, semi-natural and shrub, and assumed unchannelled flow condition for bare soil, rice paddy, and orchard using values presented in Morgan and Duzant [40]. d_a values for other LULCs derived from furrow heights of the fields, using field measurement data for dry crops and data from the Rural Development Administration of South Korea [48] for ginseng; [j] According to the guide values for Manning's *n* from Morgan [49], the values of *n* for natural land covers (i.e., forest, semi-natural, and shrub), crop fields, ginseng, and smooth surfaces (bare soil and urban) are 0.2, 0.1, 0.2, and 0.01, referring to natural range land, average tillage conditions, wheat mulching, and smooth bare soil or asphalt conditions, respectively; * The permeable black awning screen is generally installed 1.3 m above the ginseng field [48], and it acts as a plant canopy. Therefore, the cover ratio of the screen in the field and height of the screen is utilized for canopy cover (*CC*) and plant height (*PH*) values for ginseng.

2.4. Model Calibration and Validation

The DMMF model was calibrated and validated for stream discharge and suspended sediment to test its performance in the Haean catchment. The testing was performed utilizing data from the year 2010 when the LULC map, as well as the field-measured stream discharge and suspended sediment data, were well established [30,35]. We confined the testing period from the 67th day of the year (DOY), which is three days after first heavy rainfall of the year, to reduce the uncertainty of initial soil water content by equating it with the soil water content at field capacity. We equalized the two parameters based on the field measurement guidelines for soil water content at field capacity, which recommend soil sampling two or three days after rainfall that is heavy enough to saturate the soil. The three sub-catchments of S1, S2, and S3 (see Figure 1) were selected for model calibration and validation. The data from the S1 and S2 were utilized for two-step calibration, and those from the S3 were used for model validation. Two-step calibration was performed on the forest-related parameters utilizing the data from the S1 site, and the other parameters were calibrated utilizing the data from S2. This calibration method enables us to prevent the significance of forest-related parameters of dominant LULC type in the entire catchment, from overtaking the importance of other parameters, resulting in those parameters being ignored. The DMMF model can estimate the outputs of the surface and subsurface runoff, and the sediment from the elements. However, the measured data are stream discharge and suspended sediments at the outlet of each sub-catchment. Because the model does not consider in-stream processes and the impact of groundwater on the base flow of the stream, it is not appropriate to directly compare the result from the model with the measured data. To match different comparative objects, we compared the total daily discharge of each site to total daily surface runoff and subsurface interflow flowing into the stream from the model, while adding a constant corresponding to base flow from groundwater. To match the sediment yield from the terrestrial part with the suspended sediments measured at the outlet of each sub-catchment, we should consider the in-stream sediment processes and impact of erosion control facilities. Reflecting sediment deposition on the stream bed load, we assumed that only a part of the terrestrial sediment yield entering the stream was sampled at each measuring point for each sub-catchment. Therefore, we compared the suspended sediments measured from the outlet of each measuring point to the sediment flowing into the stream from the model, multiplied by a constant, reflecting the in-stream sediment process. Our assumptions can be described as below,

$$Q_m = Q_s + IF_s + \alpha, \tag{1}$$

$$SL_m = \beta \times SL_s. \tag{2}$$

Here, Q_m represents the measured daily total discharge, and Q_s, IF_s, and α represent the daily surface runoff, daily subsurface interflow simulated from the DMMF model, and a constant reflecting the base flow from groundwater (unit: m^3/s). SL_m represents the total daily suspended sediments measured at the outlet of each sub-catchment, and SL_s and β represent the terrestrial sediment yield entering the stream from the model simulation and constant representing the in-stream sediment deposition rate, respectively.

2.4.1. Sensitivity Analysis

To select important parameters to be calibrated among unmeasured or highly uncertain parameters, we performed site-specific sensitivity analyses, using the Sobol' method [50–52]. The Sobol' method is a variance based sensitivity analysis that is widely used in environmental and hydrological modeling, such as SWAT and TOPMODEL [53,54]. This method can estimate the total effect of each parameter on the model output, considering the combined effects among parameters. Therefore, the Sobol' method is more suitable for analyzing the sensitivity of non-linear and non-additive models containing many parameters, as opposed to the local or one-at-a-time (OAT) methods [53,55]. The relative sensitivity of parameters is expressed as the Sobol' total index (*SI*)—the ratio of the amount of total variance caused by a parameter to the amount of variance induced from all parameters (i.e., the unconditional variance of the model) [52]. If we have p-dimensional parameter set, the first-order sensitivity of the *i*-th parameter can be described as,

$$S_i = \frac{V_{X_i}(E_{X_{-i}}(Y|X_i))}{V(Y)}, \tag{3}$$

where $V_{X_{-i}}(E_{X_i}(Y|X_{-i}))$ is the variance of the model solely by *i*-th parameter (X_i). Then the total sensitivity of the *i*-th parameter (SI_i) can be calculated as below,

$$SI_i = 1 - \frac{V_{X_{-i}}(E_{X_i}(Y|X_{-i}))}{V(Y)}, \tag{4}$$

where $\dfrac{V_{X_{-i}}(E_{X_i}(Y|X_{-i}))}{V(Y)}$ indicates that the sum of first-order sensitivities of all parameters except *i*-th parameter. Parameters with large *SI* indicate a relatively high impact on the model output, while those with small *SI* indicate a relatively low impact on the model output.

Because the soil hydraulic parameters (i.e., θ_{sat}, θ_{fc}, and K), soil detachabilities (i.e., DK_c, DK_z, DK_s, DR_c, DR_z, and DR_s) and LULC parameters (i.e., PI, IMP, GC, CC, PH, D, NV, d, and n) were not measured or had high uncertainties, their importance was tested on model outputs. Before performing sensitivity analysis, we set the range of the parameters to be tested. The ranges of soil hydraulic parameters (i.e., θ_{sat}, θ_{fc}, and K) were set based on the range of estimated values for each represented soil class (see Table 2). The upper bound of θ_{fc} was set as the minimum θ_{sat}, and the upper bound of K was set to 18 times of the maximum K to reflect high uncertainty of the parameter [56]. The ranges of the un-measured LULC parameters were set based on the initial parameter settings for each LULC type (see Table 3). We adjusted the parameters using a range of $\pm 100\%$ for the initial parameter settings for each LULC type. If the upper or lower limits of the proportional parameters is out of the range between zero to one, we set the lower limits to zero and the upper limits to one. In this study, SIs for the input parameters were estimated using the "sobolmartinez" function of the "sensitivity" package [57] on R version 3.5.1 [58], a well-established open-source program for statistical computing, providing many analysis packages. We used the default bootstrapping option of the function, employing a sample size of 10^3.

2.4.2. Calibration

To find the optimal combination of the parameter set, which allows model outputs to explain the measured stream discharge and suspended sediments from each site, we performed two-step calibration. For each step, we adjusted the important parameters with SI greater than 0.05 (i.e., contributing 5% of the total variance), and we adjusted the constants for the in-stream processes (α and β) additionally for sub-catchment S2, where data were measured in the stream outlet. We searched for the optimal combination of the parameter set, using the differential evolution (DE) optimization method [59,60]. The DE algorithm is a heuristic optimization method with an evolution strategy for finding the global optimum value. Requiring few prerequisites for its execution, the algorithm is applicable to non-differential, nonlinear, and multimodal models. As a result, the DE algorithm has been applied to a variety of fields including hydrological model calibration [15,59–63]. We applied the DE algorithm for model calibration using the "DEoptim" package [61,64] on R version 3.5.1 [58]. We used the Nash-Sutcliffe efficiency coefficient (NSE) [65] between model outputs and field-measured data as an objective function for the DE algorithm. To treat NSE values from stream discharge and suspended sediments fairly, we evaluated the NSE values for each measurement and used the average NSE value as the final objective function:

$$F_{obj} = 1 - \frac{\mathrm{NSE}(Q_m) + \mathrm{NSE}(SL_m)}{2}, \tag{5}$$

where F_{obj} is the objective function to evaluate the model performance. We ran the function for 10^3 iterations, and ran for three different initial states to try to find the global minimum as an optimum value.

2.4.3. Validation

Using adjusted parameters from calibration steps, model performance was tested for the S3 site, which is located near the catchment outlet. Considering site-specific base flow from groundwater and in-stream sediment processes for the S3 site, we adjusted the constants for the in-stream processes (α and β). We utilized the NSE, the percent bias (PBIAS), and the coefficient of determination (R^2) as statistical criteria for model performance evaluation [66,67]. The function "gof" from the "hydroGOF" package [68] in R version 3.5.1 [58] was used to evaluate statistical criteria.

2.5. Identifying Annual Sediment Redistribution Patterns and Assessing Soil Erosion Risk

Projecting validated parameters on the DMMF model, we simulated and calculated the annual sediment redistribution patterns of the catchment. Based on the simulated result, we assessed the net soil erosion rate (SL_{net}: t/ha/year) for each element of the catchment. SL_{net} is the net soil erosion for each element, which is the amount of sediment input to each element from upslope elements (SL_{in}) subtracted from the amount of sediment output from the element (SL_{out}). Soil erosion risk was assessed by using SL_{net} of each element. We classified SL_{net} into five categories, namely tolerable, low, moderate, high, and severe, as shown in Table 4 according to the soil erosion risk categories defined by OECD [69,70] which is one of the internationally used criteria.

Table 4. Soil erosion risk categories defined by OECD [69,70].

Erosion Class	Tolerable	Low	Moderate	High	Severe
Soil erosion rate (t/ha/year)	<6	6–10.9	11–21.9	22–32.9	>33

Based on the net soil erosion rate of the entire catchment, we assessed the soil erosion characteristics for each LULC class. For the assessment, we calculated the mean SL_{net} for each LULC class.

2.6. Evaluation of the Impact of Spatial Reconfiguration of Erosion Hot Spots into Forest

We assessed the impact of the spatial reconfiguration of erosion hot spots into forest, based on the annual sediment redistribution patterns of the catchment. Erosion hot spots represent elements in which much annual net soil erosion (SL_{net}) occurs. To compare the impact of spatial reconfiguration, we calculated the annual sediment yields being generated from the terrestrial area and entering to the water bodies of the entire catchment (SY_{base}) as a base line condition. SY_{base} is the total amount of sediment yields entering the water bodies of the entire catchment, which is equal to the total amount of SL_{in} flowing into water bodies. To increase the robustness of our analysis, we only used the values between the 2.5th percentile and the 97.5th percentile for all the elements in the catchment to exclude the impact of extreme values that can occur from model outputs. The lower extreme values were set to the value of the 2.5th percentile and the upper extreme values were set to the value of the 97.5th percentile. The impact of the spatial reconfiguration of erosion hot spots into forest was evaluated by calculating the total annual sediment yields entering the stream (SY_{tot}), using the DMMF model as bare soil and croplands (i.e., bean, cabbage, ginseng, orchard, potato, radish and rice field) being sequentially changed into the forest. We selected forest, the original LULC type before anthropogenic land cover changes, as the alternative LULC to mitigate erosion-prone areas. Similar to the methods Chaplin-Kramer et al. [23] and Chaplin-Kramer et al. [24] which compute ecosystem services by marginally changing forest into agricultural areas, we computed SY_{tot} by gradually converting 1% of the bare soil and croplands in the catchment into forest until all bare soil and croplands elements are converted into forest. Based on this result, we presented the total sediment yields (SY_{tot}), reduction rate of the sediment yields entering the stream compared to base line condition (SY_{base}), and sediment yield reduction efficiency per conversion area (t/m^2).

3. Results

3.1. Model Performance

According to the calibration and validation results, the DMMF model showed competitive performance, predicting stream discharge, but showed poorer performance in evaluating the amount of suspended sediments at the outlet of each sub-catchment. We performed two-step calibration by comparing the model outputs to the measured data collected from sub-catchment S1 and S2. The LULC and soil types of sub-catchment S1 are classified as forest and forest soil, according to Tables 2 and 3. The calculated Sobol' index for important parameters, both for stream discharge (SI_Q) and suspended sediments to the stream (SI_{SL}), are presented in Table 5.

Table 5. List of important parameters from forested site with Sobol' index greater than 0.05 for stream discharge (SI_Q) and suspended sediment to the stream (SI_{SL}), and their optimized values from the DE algorithm.

Parameters	Soil Class/LULC	SI_Q	SI_{SL}	Optimized Values
θ_{fc}	Forest soil	0.035	0.118	2.24×10^{-1}
K	Forest soil	0.202	0.082	6.17×10^{1}
DR_c	Forest soil	0	0.213	2.25×10^{-1}
PI	Forest	0.781	0.180	6.66×10^{-5}
GC	Forest	0	0.775	9.92×10^{-1}
d_a	Forest	0	0.144	7.77×10^{-3}

According to the Sobol' index, the amount of stream discharge was highly influenced by the permanent interception of rainfall (PI) and lateral soil hydraulic conductivity (K), which regulate the amount of rainfall and flow rate of subsurface interflow of the sub-catchment, respectively. Vegetation and surface cover structures (GC, PI, and d_a), detachability of clay particles (DR_c), soil water content at field capacity (θ_{fc}), and lateral soil hydraulic conductivity (K) exhibited a relatively large impact on suspended sediments generated from the sub-catchment. This result indicates that the suspended sediments generated from the sub-catchment are determined by the amount of surface runoff and the erosivity of surface, because PI, K, and θ_{fc} determine the amount of surface runoff by regulating the amount of rainfall and partitioning the rate of surface and subsurface water. Parameters GC, d_a, and DR_c determine the erosivity by surface runoff.

We determined an optimized parameter set by adjusting selected important parameters from sensitivity analysis using the DE algorithm (see Table 5). With the optimized parameter set, the stream discharge and suspended sediment from the model outputs were compared with those from field measurements (see Figure 6).

After calibrating the forest-related parameters, we calibrated the other parameters, based on the measurement data collected from sub-catchment S2. We calculated the relative importance of parameters for both the stream discharge (SI_Q) and suspended sediments to the stream (SI_{SL}), using the Sobol' index, and presented them in Table 6.

Table 6. List of important parameters ($SI > 0.05$) for stream discharge (SI_Q) and suspended sediment (SI_{SL}), and their optimized values from DE algorithm.

Parameters	Soil Class/LULC	SI_Q	SI_{SL}	Optimized Values
θ_{fc}	Moderate to steep dry field soil	0.115	0.112	3.18×10^{-1}
K	Moderate to steep dry field soil	0.223	0.020	6.06×10^{-1}
K	Flat dry field soil	0.062	0.001	1.59×10^{-1}
DR_c	Moderate to steep dry field soil	0	0.217	1.39
DR_z	Moderate to steep dry field soil	0	0.119	9.59×10^{-1}
PI	Semi-natural	0.252	0.048	4.16×10^{-4}
PI	Rice paddy	0.101	0.000	2.91×10^{-1}
PI	Other dry crops	0.178	0.011	1.28×10^{-4}
GC	Semi-natural	0	0.080	3.60×10^{-2}
d_a	Semi-natural	0	0.158	1.74×10^{-1}
d_a	Bean	0	0.105	2.93×10^{-1}
α	-	-	-	1.75×10^{-2}
β	-	-	-	4.57×10^{-2}

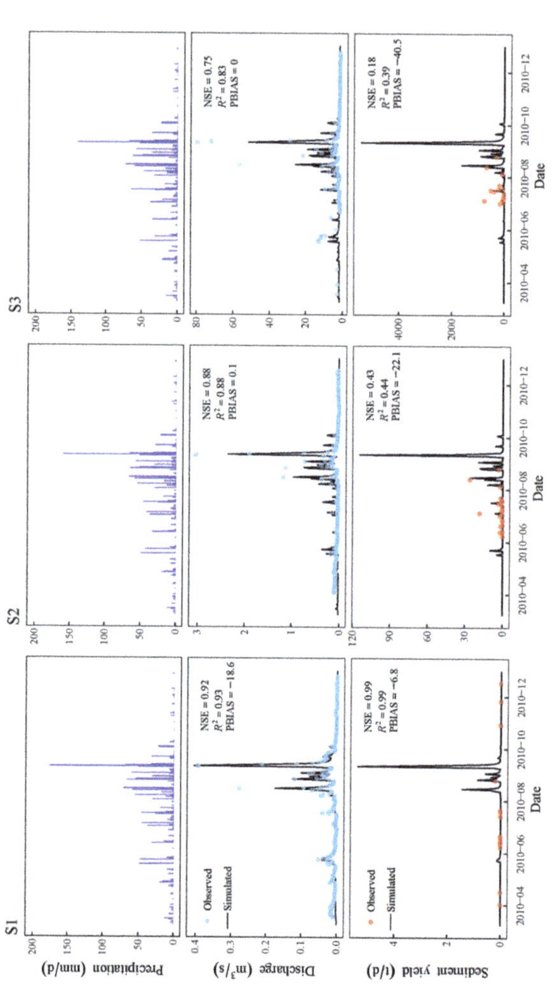

Figure 6. Calibration (S1 and S2) and validation (S3) result for stream discharge, and suspended sediment measured at the outlet of each sub-catchment.

According to sensitivity analysis, model outputs were highly sensitive to soil hydraulic characteristics of moderate to steep dry field soil and land cover structures of the semi-natural field. In details, the stream discharge of the sub-catchment was highly sensitive to the permanent interception of rainfall (PI) of the semi-natural, rice paddy, and other dry crops; the lateral hydraulic conductivity (K) of the moderate to steep dry field and flat dry field soils; and the soil water content at field capacity (θ_{fc}) of the moderate to steep dry field. This result indicates that stream discharge is highly influenced by the amount of rainfall reaching the ground (PIs) and the flow rate of subsurface interflow (Ks and θ_{fc}) of this region. The sediment yield to the stream is sensitive to the soil detachability by runoff (DR_c and DR_z) of the moderate to steep dry field soil, soil water content at field capacity (θ_{fc}) of the moderate to steep dry field soil, flow depth (d_a) of the semi-natural field and bean field, and ground cover ratio (GC) of the semi-natural field. This result emphasizes the role of the moderate to steep dry field soil, which is the second largest soil type, following forest soil, and demonstrates the crucial role of the semi-natural field on determining suspended sediment output from the model.

The performance statistics for the calibration and its time series plots of observed versus simulated stream discharge and suspended sediment were presented in Figure 6. For the calibration steps, the NSE values for stream discharge were 0.92 and 0.88 for sub-catchment S1 and S2, respectively. The R^2 values for stream discharge were 0.93 and 0.88, respectively, and the PBIAS values for stream discharge were -18.6 and 0.1, respectively. The NSE values for suspended sediment were 0.99 and 0.43 for sub-catchments S1 and S2, respectively. The R^2 values for suspended sediment were 0.99 and 0.44, and the PBIAS values for suspended sediment were -6.8 and -22.1 for the sub-catchments, respectively. The site-specific constants reflecting the baseflow from groundwater (α) and in-stream sediment deposition rate (β) for sub-catchment S2 are $1.75 \times 10^{-2}\,\mathrm{m^3/s}$ and 4.57×10^{-2}. In validation steps, the NSE values for stream discharge and suspended sediment were 0.75 and 0.18, respectively, with the site-specific α and β being $1.711\,\mathrm{m^3/s}$ and 6.76×10^{-2}, respectively. The R^2 for discharge and sediment were 0.83 and 0.39, respectively, and the PBIAS for discharge and sediment were 0 and -40.5, respectively. According to the model performance evaluation criteria suggested by Moriasi et al. [67], the DMMF model showed good performance for discharge in both calibration and validation steps. Though there is no clear model performance evaluation criteria suggested for daily time scale sediment result for watershed model due to limited reported data [67], When we apply the performance evaluation criteria for monthly time scale sediment result for watershed scale model, the model might be considered to have a slightly poor performance for sediment during the calibration and validation steps, as the NSE and R^2 values were less than 0.45 and 0.40, respectively.

3.2. Sediment Redistribution Pattern of the Catchment

Simulating the model with optimized parameters, we calculated the annual net soil erosion rate (SL_{net}) for each element and classified them into five classes–tolerate, low, moderate, high, and severe—as in Figure 7.

According to Figure 7, elements with severe soil erosion ($>33\,\mathrm{t/ha/year}$) were concentrated on the dry crop field with moderate to steep slope conditions on the interface with the forest. The estimate of the mean annual net soil erosion rate by each LULC type (Table 7) shows that bare soil and dry crop field suffered from severe soil erosion. On the other hand, forest, rice paddy, orchard, and urban areas showed good sediment capturing capabilities.

(a)

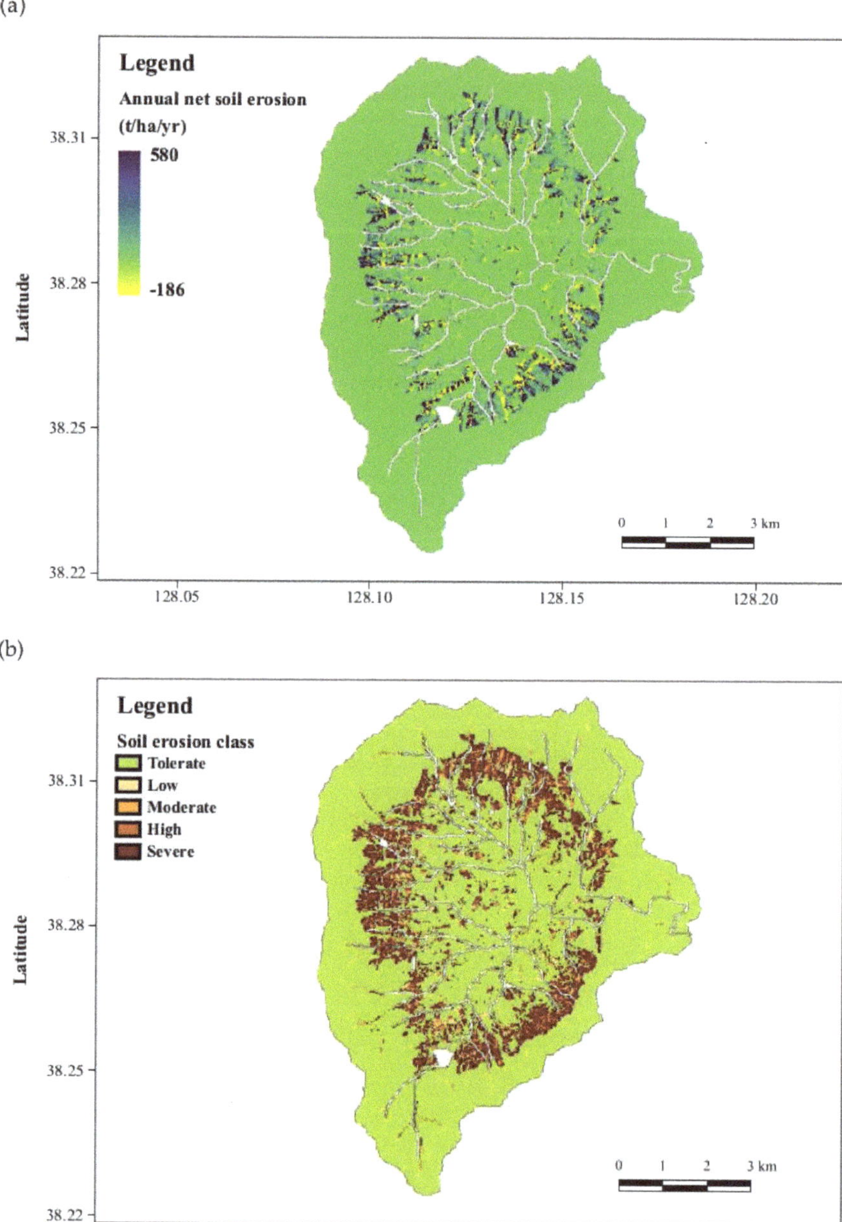

(b)

Figure 7. (**a**) Annual net soil erosion (t/ha/year) of the entire Haean catchment and (**b**) soil erosion class according to the soil erosion risk categories from OECD [69,70].

Table 7. Mean annual net soil erosion rate (t/ha/year) and mean slope of each LULC type.

LULC	Mean Annual Net Soil Erosion Rate (t/ha/year)	Mean Slope (°)
Bare soil	997.80	9.8
Bean	763.82	7.6
Ginseng	388.83	8.5
Potato	357.60	7.9
Radish	310.06	8.4
Other dry crops	294.23	8.4
Semi-natural	126.34	9.0
Shrub	105.54	11.1
Cabbage	79.30	7.6
Catchment average	52.68	16.0
Forest	−75.25	22.0
Rice paddy	−171.83	3.0
Orchard	−227.14	8.1
Urban	−284.71	6.0

3.3. Impacts of Conversion of Erosion Hot Spots into Forest on Total Sediment Yield Entering the Stream

The LULC conversion of erosion hot spots into forest showed a dramatic impact in the reduction of sediment yields entering the stream, as shown in Figure 8.

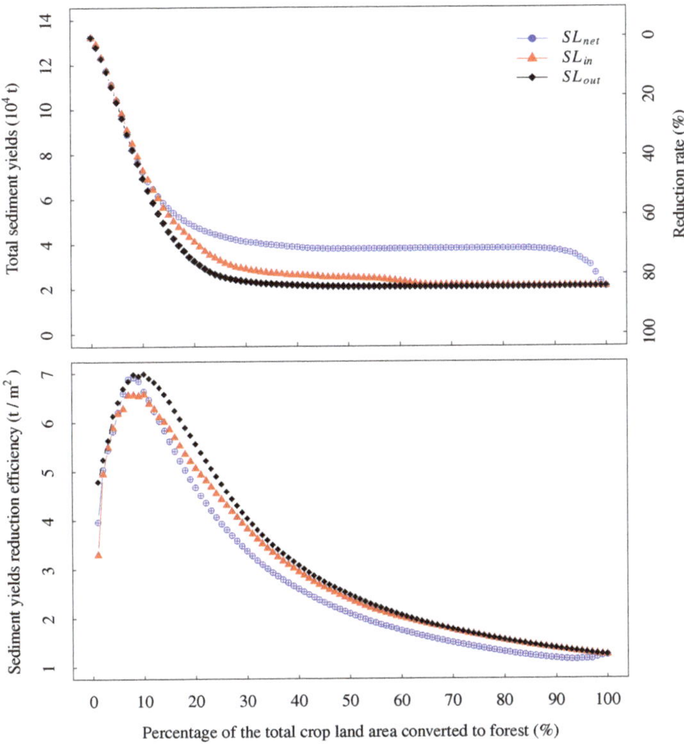

Figure 8. Total annual sediment yields entering the stream (**upper panel**) and sediment yield reduction efficiency per unit conversion area (**lower panel**) through changing bare soil and crop fields into forest sequentially from the area with the highest to the area with the lowest amount of net soil erosion (SL_{net}), sediment inflow to the element (SL_{in}), and sediment output from the element (SL_{out}).

When each bare soil and crop field element in the catchment was converted into the forest sequentially from the area with the highest soil erosion rate to the area with the lowest soil erosion rate, the amount of total annual sediment yield of the catchment to the stream sharply decreased having a shape similar to an inverted sigmoid function. Changing the 3% of erosion hot spots that have suffered the most from severe soil erosion caused a reduction in sediment yield entering the stream of ca. 10% from the baseline condition (SY_{base}), and a change in 10% of most severe hot spots is expected to reduce sediment yields by ca. 50%. Among the elements SL_{net}, SL_{in}, and SL_{out}, the altered areas revealed that outputs from the element (SL_{out}) proved to be the most effective in reducing the total sediment yield into the stream. A simulation of the sediment yields entering the stream showed that the reducing rate in sediment yield for SL_{net} was less effective than those for SL_{out} and SL_{in}. Due to total annual sediment yields sigmoidally decreases as bare soil and crop fields begin changed into forest, sediment yield reduction efficiency per unit conversion area increased until ca. 10% of total crop land area converted to forest and then gradually decreased. A simulation of the sediment yield reduction efficiency showed that the element (SL_{out}) was most efficient for all conversion intervals.

4. Discussion

Our findings emphasize the importance of landscape configuration on regulating ecosystem services by showing the effectiveness of spatial reconfiguration of soil erosion hot spots into forest on reducing the amount of sediment yield entering the stream. We simulated the annual sediment redistribution pattern in the Haean catchment, utilizing the daily based Morgan–Morgan–Finney (DMMF) soil erosion model. According to the result, the soil erosion rate varied greatly depending on the topography and LULC type, and the area located on the steep hill-slope, which is adjacent to the forest severely suffered from soil erosion. When reconfiguring the landscape patterns of croplands by sequentially altering erosion hot spots from the most severe to the least severe areas into forest, we found dramatic effects in the reduction of sediment yields entering the stream in this catchment. The reduction rate may reach ca. 50% when the 10% most severe erosion hot spots were altered, and we can expect a reduction rate of over 80% when the ca. 20% most severe erosion hot spots are altered. In the following, we first discuss model performance and limitation, and then potential management implications.

4.1. Model Performance

The assessment of soil erosion risk and measurement of the effectiveness of the spatial reconfiguration of erosion hot spots in reducing sediment yields entering the stream were based on the calibrated and validated simulations of the DMMF soil erosion model. According to the model performance criteria from Moriasi et al. [67], the DMMF model showed satisfactory performance for predicting stream discharge during the calibration and validation processes, with mean NSE values of 0.90 and 0.75, mean R^2 of 0.91 and 0.83, and maximum PBIAS of −18.6 and 0 during calibration and validation steps, respectively. The model showed comparatively poor performance for predicting suspended sediment at the outlet of each sub-catchment, except the small forested site (S1) where the stream does not exist. The mean NSE values were 0.66 and 0.18, mean R^2 were 0.67 and 0.39, and maximum PBIAS were −22.1 and −40.5, respectively. When we compared the model performance statistics of the DMMF model to those from previous studies using soil and water analysis tool (SWAT), the model showed competitive performance in predicting stream discharge but poorer performance in terms of predicting suspended sediments in the stream [3,19]. Maharjan et al. [3] reported that mean NSE values for stream discharge were 0.82 during calibration and 0.45 during validation. In addition, they showed that mean NSE values for suspended sediment were 0.78 and 0.60 during calibration and validation, respectively. Jang et al. [19] also reported mean NSE values for stream discharge of 0.78 and 0.66 during calibration and validation, respectively. They reported mean R^2 for suspended sediment were 0.80 and 0.76 during calibration and validation, respectively. In terms of soil erosion rate for each crop field, the DMMF model estimated that the average annual soil loss of major dry crops ranged between 79.3 t/ha/year and 763.8 t/ha/year for bean, radish, potato, and cabbage, and the average annual soil loss from whole dry crop fields was 379.7 t/ha/year. Arnhold et al. [17] reported that 30–54 t/ha/year of soil loss occurred in the dry crop fields, including bean, radish, potato, and cabbage, from the plot-level field measurement. Furthermore, Maharjan et al. [3] estimated that 35.5–53.0 t/ha/year of soil loss occurred in the dry crop fields from the SWAT model. When we compared the results from the DMMF model with those from other studies, the amount of soil loss from this study is far greater. The reasons that the DMMF model showed poor performance for predicting suspended sediment in the stream can be analyzed from two perspectives. The first reason involves the discrepancy of data types between the DMMF model and observed data. The observed data were stream discharge and suspended sediment at the outlet of each sub-catchment. On the other hand, the DMMF model can estimate the total sediment yields entering the stream that belongs to each sub-catchment. The DMMF model is efficient for estimating sheet and rill erosion, but it has limitations in estimating in-stream sediment processes such as stream bed deposition, channel erosion, and sediment transport in the stream. Considering the limitations of the model, we use site-specific coefficients, which assume that suspended sediments measured at the outlet are proportional to the sediment yields inflowing into the stream. However, incorporating the quantity and the velocity of stream water discharge, sediment flux, and physical characteristics of channel structures such as gradient, width, depth, and length, into the in-stream sediment process, is complicated [71,72]. The reasons above may lead to a high sediment deposition rate in the stream (i.e., low measured sediment ratio (β)), which in turn, causes a high soil erosion rate in the terrestrial area. Because the study sites are affected by monsoon climate, such that its rainfall pattern is not uniform but rather with a lot of extremes, a large amount of sediment is deposited during low rainfall events, and the deposited sediments are washed out by a huge amount of fast stream discharge accompanying heavy rainfall. Temporal lags between the rainfall event and stream discharge are negligible for the Haean catchment, but for suspended sediments, the lags are significant and highly depend on the stream length because of the difference in travel velocities between water and soil particles [73–76]. Therefore, the model performance for predicting stream discharge may be better than that for predicting suspended sediments. The stream widens and deepens as it descends to the lower area, according to Lee [2], and the length of the stream also increases as the size of sub-catchment grows. The uncertainty caused by in-stream processes increases as the size of the sub-catchment grows, which reduces the model performance in predicting suspended sediments in this study. SWAT and USLE-based models are usually calibrated and validated at the fixed spatial area with a different temporal period. Therefore, in-stream sediment processes can be included in the parameters, which may lead to better model performance. However, the DMMF model is a spatially distributed semi-processed model and used the same temporal period with a different spatial area for calibration and validation in this research, so that the in-stream processes cannot be included in the model.

Secondly, many sediment reduction facilities, such as dams for freshwater, debris barrier and culvert systems around crop fields, and road infrastructures, which can affect sediment transport processes, have been installed in the Haean catchment [30,36]. The dam and debris barriers create reservoirs that impede the stream flow and filter out sediments in the facilities. This disrupts the correct evaluation of the model performance for this catchment. Shope et al. [30] showed complex stream networks, including the culvert systems around crop fields and the road infrastructure. The culvert systems extend the travel time of suspended sediments and reduce the runoff and transport velocities of sediments by altering the flow direction abruptly. Increased travel time and decreased transport velocity tend to increase the deposition rate of sediments compared to the condition without the culvert system. The deposited sediments in the culvert flow into the stream by runoff, with sufficient power to wash out. The culvert system is also responsible for the temporal lag between the rainfall event and the presence of suspended sediments in the catchment. Sediment reduction facilities trap a huge amount of sediments, which make the measured sediment ratio (β) in this study have very low values. Because of the small β, the stream bed deposition rate became too large, and consequently, the overall erosion rate from terrestrial area increased. To cope with this problem, the in-stream processes will need to be considered more precisely through model improvements.

4.2. Assessment of Soil Erosion Risk and the Effectiveness of Spatial Reconfiguration of Erosion Hot Spots on Reducing Sediment Yield Entering the Stream

We estimated the annual net soil erosion rate of the entire catchment and assessed the soil erosion risk class according to the OECD criteria. According to this study, soil erosion is concentrated on the hill-slope of the catchment, and the problem is more significant for the bare soil and dry crop fields, such as bean, radish, and potato, in this area. In addition, forest in the valley showed a considerable amount of soil loss, also suffering from erosion due to the concentrated surface runoff and steep slope. Compared with other studies, the soil erosion risk pattern and the average annual soil loss from the DMMF model is qualitatively consistent with the soil erosion risk map from Lee et al. [77], with average climate conditions for the 2010s using the USLE-based SATEEC [78] model. According to this study, urban area, orchard, and rice field showed better performance for sediment capturing capabilities than forest. However, the urban area and rice field are located in the lower and flatter area than forest, so that the sediment inputs from the upslope area tend to be deposited in this area. Furthermore, because the urban area is usually paved with impervious covers, such as concrete and asphalt, and the rice field is filled with water, which acts as a pervious cover that prevents surface erosion, these areas have little soil loss but receive huge input from the upslope area. Though the forest is in a region where the slope is very steep, the average amount of soil loss is smaller compared with other land types, and it also shows excellent sediment capturing capability, in general. Like the other studies, we can conclude that the main cause of severe erosion in the catchment is cropland extension after deforestation at the hill-slope area of the catchment [2,3,11,17,19,77]. We also assessed the effect of spatial reconfiguration of LULCs on reducing sediment yields entering the stream. In this study, the spatial reconfiguration of erosion hot spots into forest showed excellent reduction efficiency in sediment yields entering the stream. We identified that the sediment yields entering the stream were reduced sharply, as crop lands were sequentially changed into forest from the area with the most severe soil loss to the area with the least soil loss. An sigmoidal sediment reduction rate from altering LULCs to forest indicates that forest is not only effective in preventing surface erosion but also effective in capturing sediment input from the upslope area. In addition, the result suggested that altering LULCs based on the amount of sediment output from the element is the most effective way of reducing sediment yields entering the stream. This result is consistent with previous studies that emphasize the effectiveness of vegetative filter strips located at sediment sources such as crop fields [3,5,19,27–29]. The result can also be generalized to consider the effect of riparian vegetation buffer strip on reducing sediment yields entering the stream, located at the interface between crop fields or natural sediment sources and the stream channel [4,79,80]. This study also demonstrated that the sediment yield reduction efficiency initially increased as the first few bare soil area and crop lands with the most severe soil loss were converted into forest. The sediment yield reduction efficiency were maximized when ca. 10% of the area converted, and then the efficiency decreased gradually. These patterns can be explained by two aspects of the forest's sediment yield reduction capability; protecting surface from soil erosion, and capturing sediment inputs. The areas with the most severe soil loss are located at the steep hillslope where surface runoff is concentrated. These areas have a large transport capacity of the runoff, beyond the sediment capturing capability of forest because transport capacity is greater than the available sediment for transport [15]. In these areas, conversion of crop lands into forest can reduce soil loss from the surface but cannot capture sediment inputs from upslope which is larger than surface soil loss. As slope becomes milder and the amount of surface runoff decreases due to gradual conversion of crop lands into forest, transport capacity gradually decreases. Decreased transport capacity caused by decreased slope gradient and surface runoff lets forest capture more sediments, maintaining the surface protecting capability from soil loss. Therefore, the sediment yield reduction capabilities of forest become small and the sediment yield reduction efficiency by changing crop lands into forest decreases gradually. According to these results, one can reduce sediment yields entering stream efficiently by identifying an optimal percentage of crop land conversion into forest which brings out the best efficiency of sediment yield reduction per unit conversion area.

5. Conclusions

In this study, we identified the soil erosion risk of Haean catchment spatially explicitly by projecting sediment redistribution patterns using the DMMF model. In addition, we measured the sediment yield reduction efficiency entering the stream by sequentially altering erosion hot spots into forest from that which has the highest soil loss to that which has the lowest soil loss. The DMMF model showed competitive performance estimating stream discharge but exhibited lower performance estimating suspended sediments at each sub-catchment outlet. When we applied the DMMF model to the Haean catchment, the bare soil surface and dry crop fields located on the steep hill-slope of the catchment suffered mostly from severe soil erosion. On the other hand, forest, rice paddy, orchard, and urban areas suffer less from soil erosion. By changing the erosion hot spots from cropland to forest, the overall amount of sediments exporting to the stream of the catchment was effectively reduced. The sediment yield reduction efficiency was maximized when ca. 10% of crop lands were converted to forest. This study implies that one can achieve the goal of reducing sediment yields entering the stream by identifying the location of erosion hot spots and managing the area intensively. Although previous studies showed good mitigation effects of BMPs that require compliance of stakeholders, this may not be easy and takes much time for stakeholders to follow the BMPs, because the degree of acceptance of the policy depends on the situation and tendency of each stakeholder [19]. On the other hand, the spatial reconfiguration approach proposed in this study can reduce the number of stakeholders relevant to soil erosion mitigation measures. However, this approach reduces crop yields because crop lands are converted to non-crop lands to reduce sediment yields from the catchment. In addition, the sediment yield reduction efficiency decreases after a certain point of spatial reconfiguration. Therefore, the two approaches—BMP measures such as cultivating cover crops, mulching surface with straw, and managing field margin naturally, and conversion of crop lands with the more severe soil loss—are complementary measures to reduce sediment yields into the stream.

Author Contributions: Conceptualization, K.C. and B.R.; Data curation, K.C. and G.R.M.; Formal analysis, K.C.; Investigation, K.C.; Methodology, K.C.; Resources, G.R.M.; Software, K.C.; Supervision, B.R.; Validation, K.C.; Visualization, K.C.; Writing original draft, K.C.; Writing review & editing, G.R.M. and B.R.

Funding: This study is part of the International Research Training Group "Complex Terrain and Ecological Heterogeneity" (TERRECO) funded by the German Research Foundation (DFG) with the grant number [GRK 1565/1]. This study was also supported by the National Research Foundation of Korea (NRF) with the grant number [NRF-2017R1A2B4010460].

Acknowledgments: This paper is dedicated to the memory of our wonderful colleague, Sebastian Arnhold, who passed away last year. This publication was funded by the German Research Foundation (DFG), the National Research Foundation of Korea (NRF). This publication was also funded by the German Research Foundation (DFG) and the University of Bayreuth in the funding programme Open Access Publishing. We would like to thank Editage (www.editage.co.kr) for English language editing.

Appendix A. Detailed Structure of the Daily Based Morgan–Morgan–Finney (DMMF) Soil Erosion Model

Morgan–Morgan–Finney (MMF) model [37] is a conceptual soil erosion model, which estimates the annual soil erosion rate from an area by comparing the amount soil particles detached from the surface (SS) and transport capacity of surface runoff (TC) [37,38,40]. The first version of MMF model [37] estimated soil erosion rate of an area by comparing the amount of soil particles detached by raindrop impact (F) and transport capacity of surface runoff (TC). The second version of model, the revised Morgan–Morgan–Finney (RMMF) model [38] started to consider the amount of soil particles generated by surface runoff (H). In the third version, the modified Morgan–Morgan–Finney (MMMF) model [40], the interconnectivity of surface runoff, various sub-processes such as the subsurface interflow and gravitational deposition processes, and parameters such as the physical structure of vegetation and surface ground conditions were introduced to calculate transport capacity of surface runoff (TC) and the amount of soil particles available for transport (G) more physically rigorously [41]. The daily based Morgan–Morgan–Finney (DMMF) soil erosion model [15] is also estimates daily soil loss from an element by comparing transport capacity of surface runoff (TC) and the available sediment for transport (G). The DMMF model is mainly comprised of hydrological and sediment phases. The hydrological phase determines the amount of surface runoff and subsurface interflow, and the sediment phase determines the amount of sediment budgets of the element.

Appendix A.1. Hydrological Phase

The effective rainfall (R_{eff}; mm) which is the volume of rainfall reaching the unit surface area of an element is the main driver of hydrological phase. Following the corrected version of the effective rainfall (R_{eff}) from Choi et al. [42], R_{eff} is calculated as,

$$R_{eff} = R \times (1 - PI) \times \cos(S), \tag{A1}$$

where PI is the proportion of the permanent interception area and S is the slope of an element. Similar to MMF model, surface runoff can be generated when the total input of water to the element exceeds the surface water infiltration capacity (SW_c; mm), which is the soil moisture storage capacity considering the proportion of the impervious area (IMP). SW_c is defined as,

$$SW_c = (1 - IMP) \times (SW_{sat} - SW_{init} - \frac{\Sigma IF_{in}}{A}), \tag{A2}$$

where SW_{sat} (mm) is the volume of water per unit area when soil is fully saturated, and SW_{init} (mm) is the volume of initial water per unit area that is already existed in the soil. ΣIF_{in} (L) is the volume of subsurface water inputs from upslope and A (m^2) is the area of an element. The amount of the surface runoff (Q; mm) is calculated as,

$$Q = R_{eff} + \frac{\Sigma Q_{in}}{A} - SW_c, \tag{A3}$$

where Q_{in} (L is the volume of surface runoff inflow from upslope areas. The amount of water in the soil also flows out from the element as a subsurface interflow (IF_{out}; L) when the voludme of soil water budget per unit area (SW; mm) of the element exceeds the volume of soil water at field capacity per unit area (SW_{fc}; mm). The soil water budget (SW) is estimated as,

$$SW = (SW_{init} + \frac{\Sigma IF_{in}}{A}) + (R_{eff} + \frac{\Sigma Q_{in}}{A} - Q) - ET, \tag{A4}$$

where ET (mm) is the volume of water evapotranspirates per unit area from the element. Then the volume of subsurface water flowing out from the element (IF_{out}) can be described as,

$$IF_{out} = K \times \sin(S) \times (SW - SW_{fc}) \times w, \tag{A5}$$

where K (m/day) is the saturated soil lateral hydraulic conductivity and w (m) is the width of the element. A part of soil water remains with remaining water content (θ_r; vol/vol) which can be described as,

$$\theta_r = \frac{(SW - IF_{out}/A)}{1000 \times SD}, \tag{A6}$$

where SD is the soil depth of the element, and 1000 is the constant to convert meters to millimeters. The θ_r can be changed into θ_{init} for the next day.

Appendix A.2. Sediment Phase

Sediment phase determines the total mass of soil particles which is taken out of the element through three steps: delivery of detached soil particles into the surface runoff, gravitational deposition, and estimation of hhe sediment loss from the element (SL) by comparing transport capacity of the runoff (TC; kg/m^2) and sediment available for tranport (G; kg/m^2). In the model, soil particles are detached from the surface by raindrop impact and surface runoff. The mass of soil particles detached by raindrops per unit area (F; kg/m^2) is described as,

$$F = 0.001 \times DK \times P \times (1 - EPA) \times KE, \tag{A7}$$

where DK (g/J) is the detachability of soil particles by raindrop impact, P (%) is the proportion of each soil particle size class (i.e., clay, silt, and sand), KE (J/m^2) is the kinetic energy of the effective rainfall considering direct throughfall and leaf drainage from the plant, and 0.001 is the unit conversion factor from g to kg. Also, EPA is the erosion protected area:

$$EPA = IMP + (1 - IMP) \times GC, \tag{A8}$$

where GC is the proportion of ground cover and IMP is the proportion of the impervious area (IMP) of the element. The mass of detached soil particles by the surface runoff (H; kg/m^2) is described as,

$$H = 0.001 \times DR \times P \times Q^{1.5} \times (1 - EPA) \times (\sin(S))^{0.3}, \tag{A9}$$

where DR (g/mm) is the detachability of soil particles by runoff per unit volume of surface runoff and Q is the volume of runoff per unit area, S is the slope of the element, and 0.001 is the unit conversion factor from g to kg. Sediment inputs from upslope elements (ΣSL_{in}) also flows into surface runoff. The mass of delivered sediments to the surface runoff per unit area (SS; kg/m^2) is,

$$SS = F + H + \frac{\Sigma SL_{in}}{A}. \tag{A10}$$

A part of sediments delivered to the surface runoff (SS) in the runoff settle down to the ground by gravity. The gravitational deposition rate of the suspended sediments (SS) in runoff (DEP) is,

$$DEP = 0.441 \times N_f, \tag{A11}$$

where N_f is the particle fall number which is the probabilistic ratio of falling particles [81], The N_f can be estimated as,

$$N_f = \frac{l}{v} \times \frac{v_s}{d}, \tag{A12}$$

where v (m/s) is the velocity of the surface runoff, v_s is the settling velocity of each particle size class, and d (m) is the depth of the surface runoff.

The remaining suspended sediments become available for transport per unit volume of surface runoff per unit area (G; kg/m^2) and be estimated as,

$$G = SS \times (1 - DEP). \tag{A13}$$

The part of the availabe sediments for transport (G) can flow out from the element according to the transport capacity of the runoff (TC; kg/m^2) of an element which is determined by the volume of runoff per unit area of an element (Q), the slope angle (S) and the surface conditions [40]. Due to the physical condition of surface affect runoff velocity, the tranport capacity of runoff can be described using the ratio between actual runoff velocity (v) and the reference velocity of the element (v_r; m/s) [42].

$$TC = 0.001 \times \left(\frac{v}{v_r}\right) \times Q^2 \times \sin(S).$$ (A14)

The reference velocity (v_r) is,

$$v_r = \frac{1}{n_r} \times d_r^{2/3} \times \sqrt{\tan(S)},$$ (A15)

with 0.015 for Manning's coefficient (n_r) and 0.005 for runoff depth (d_r) representing for a standard surface condition. The transport capacity of the runoff (TC) and the available sediment for transport (G) determines the amount of sediment loss from the element (SL) [40,82]. When TC is greater than G, the surface runoff washes out all the sediments available for transport, otherwise, the amount of sediment (SL) which is equal to TC can be transported from the element.

References

1. Hu, Q.; Gantzer, C.J.; Jung, P.K.; Lee, B.L. Rainfall erosivity in the Republic of Korea. *J. Soil Water Conserv.* **2000**, *55*, 115–120.
2. Lee, J.Y. Importance of hydrogeological and hydrologic studies for Haean basin in Yanggu. *J. Geol. Soc. Korea* **2009**, *45*, 405–414.
3. Maharjan, G.R.; Ruidisch, M.; Shope, C.L.; Choi, K.; Huwe, B.; Kim, S.J.; Tenhunen, J.; Arnhold, S. Assessing the effectiveness of split fertilization and cover crop cultivation in order to conserve soil and water resources and improve crop productivity. *Agric. Water Manag.* **2016**, *163*, 305–318. [CrossRef]
4. Lee, K.H.; Isenhart, T.M.; Schultz, R.C. Sediment and nutrient removal in an established multi-species riparian buffer. *J. Soil Water Conserv.* **2003**, *58*, 1–8.
5. Ali, H.E.; Reineking, B. Extensive management of field margins enhances their potential for off-site soil erosion mitigation. *J. Environ. Manag.* **2016**, *169*, 202–209. [CrossRef]
6. Jeon, J.H.; Park, C.G.; Choi, D.; Kim, T. Characteristics of Suspended Sediment Loadings under Asian Summer Monsoon Climate Using the Hydrological Simulation Program-FORTRAN. *Sustainability* **2017**, *9*, 44. [CrossRef]
7. Pimentel, D.; Harvey, C.; Resosudarmo, P.; Sinclair, K.; Kurz, D.; McNair, M.; Crist, S.; Shpritz, L.; Fitton, L.; Saffouri, R.; et al. Environmental and economic costs of soil erosion and conservation benefits. *Science* **1995**, *267*, 1117–1122. [CrossRef]
8. Pimentel, D.; Kounang, N. Ecology of soil erosion in ecosystems. *Ecosystems* **1998**, *1*, 416–426. [CrossRef]
9. Lal, R. Soil degradation by erosion. *Land Degrad. Dev.* **2001**, *12*, 519–539. [CrossRef]
10. Yoon, B.; Hyoseop, W. Sediment problems in Korea. *J. Hydraul. Eng.* **2000**, *126*, 486–491. [CrossRef]
11. Arnhold, S.; Ruidisch, M.; Bartsch, S.; Shope, C.L.; Huwe, B. Simulation of runoff patterns and soil erosion on mountainous farmland with and without plastic-covered ridge-furrow cultivation in South Korea. *Trans. ASABE* **2013**, *56*, 667–679. [CrossRef]
12. Park, J.H.; Duan, L.; Kim, B.; Mitchell, M.J.; Shibata, H. Potential effects of climate change and variability on watershed biogeochemical processes and water quality in Northeast Asia. *Environ. Int.* **2010**, *36*, 212–225. [CrossRef] [PubMed]
13. Stocker, T.F.; Qin, D.; Plattner, G.K.; Alexander, L.V.; Allen, S.K.; Bindoff, N.L.; Bréon, F.M.; Church, J.A.; Cubasch, U.; Emori, S.; et al. *Climate Change 2013: The Physical Science Basis. Contribution of Working Group I to the Fifth Assessment Report of the Intergovernmental Panel on Climate Change*; Cambridge University Press: Cambridge, UK; New York, NY, USA, 2013; pp. 33–115.
14. Chang, H. Spatial analysis of water quality trends in the Han River basin, South Korea. *Water Res.* **2008**, *42*, 3285–3304. [CrossRef] [PubMed]
15. Choi, K.; Arnhold, S.; Huwe, B.; Reineking, B. Daily Based Morgan–Morgan–Finney (DMMF) Model: A Spatially Distributed Conceptual Soil Erosion Model to Simulate Complex Soil Surface Configurations. *Water* **2017**, *9*, 278. [CrossRef]

16. Ruidisch, M.; Kettering, J.; Arnhold, S.; Huwe, B. Modeling water flow in a plastic mulched ridge cultivation system on hillslopes affected by South Korean summer monsoon. *Agric. Water Manag.* **2013**, *116*, 204–217. [CrossRef]

17. Arnhold, S.; Lindner, S.; Lee, B.; Martin, E.; Kettering, J.; Nguyen, T.T.; Koellner, T.; Ok, Y.S.; Huwe, B. Conventional and organic farming: Soil erosion and conservation potential for row crop cultivation. *Geoderma* **2014**, *219–220*, 89–105. [CrossRef]

18. Arnold, J.G.; Srinivasan, R.; Muttiah, R.S.; Williams, J.R. Large area hydrologic modeling and assessment part I: Model development. *J. Am. Water Resour. Assoc.* **1998**, *34*, 73–89. [CrossRef]

19. Jang, S.S.; Ahn, S.R.; Kim, S.J. Evaluation of executable best management practices in Haean highland agricultural catchment of South Korea using SWAT. *Agric. Water Manag.* **2017**, *180*, 224–234. [CrossRef]

20. Fujisaka, S. Learning from six reasons why farmers do not adopt innovations intended to improve sustainability of upland agriculture. *Agric. Syst.* **1994**, *46*, 409–425. [CrossRef]

21. Pannell, D.J. Social and economic challenges in the development of complex farming systems. *Agrofor. Syst.* **1999**, *45*, 395–411. [CrossRef]

22. Poppenborg, P.; Koellner, T. Do attitudes toward ecosystem services determine agricultural land use practices? An analysis of farmers' decision-making in a South Korean watershed. *Land Use Policy* **2013**, *31*, 422–429. [CrossRef]

23. Chaplin-Kramer, R.; Sharp, R.P.; Mandle, L.; Sim, S.; Johnson, J.; Butnar, I.; Milà i Canals, L.; Eichelberger, B.A.; Ramler, I.; Mueller, C.; et al. Spatial patterns of agricultural expansion determine impacts on biodiversity and carbon storage. *Proc. Natl. Acad. Sci. USA* **2015**, *112*, 7402–7407. [CrossRef]

24. Chaplin-Kramer, R.; Hamel, P.; Sharp, R.; Kowal, V.; Wolny, S.; Sim, S.; Mueller, C. Landscape configuration is the primary driver of impacts on water quality associated with agricultural expansion. *Environ. Res. Lett.* **2016**, *11*, 074012. [CrossRef]

25. Lee, T. Analyzing the Effectiveness of a Best Management Practice on Sediment Yields Using a Spatially Distributed Model. *J. Korean Geogr. Soc.* **2017**, *52*, 15–24.

26. Polasky, S.; Nelson, E.; Camm, J.; Csuti, B.; Fackler, P.; Lonsdorf, E.; Montgomery, C.; White, D.; Arthur, J.; Garber-Yonts, B.; et al. Where to put things? Spatial land management to sustain biodiversity and economic returns. *Biol. Conserv.* **2008**, *141*, 1505–1524. [CrossRef]

27. Dillaha, T.A.; Reneau, R.B.; Mostaghimi, S.; Lee, D. Vegetative Filter Strips for Agricultural Nonpoint Source Pollution Control. *Trans. ASAE* **1989**, *32*, 513–519. [CrossRef]

28. Delgado, A.N.; Periago, E.L.; Viqueira, F.D.F. Vegetated filter strips for wastewater purification: A review. *Bioresour. Technol.* **1995**, *51*, 13–22. [CrossRef]

29. Muñoz-Carpena, R.; Parsons, J.E.; Gilliam, J.W. Modeling hydrology and sediment transport in vegetative filter strips. *J. Hydrol.* **1999**, *214*, 111–129. [CrossRef]

30. Shope, C.L.; Maharjan, G.R.; Tenhunen, J.; Seo, B.; Kim, K.; Riley, J.; Arnhold, S.; Koellner, T.; Ok, Y.S.; Peiffer, S.; et al. Using the SWAT model to improve process descriptions and define hydrologic partitioning in South Korea. *Hydrol. Earth Syst. Sci.* **2014**, *18*, 539–557. [CrossRef]

31. Park, S.; Oh, C.; Jeon, S.; Jung, H.; Choi, C. Soil erosion risk in Korean watersheds, assessed using the revised universal soil loss equation. *J. Hydrol.* **2011**, *399*, 263–273. [CrossRef]

32. Bartsch, S.; Peiffer, S.; Shope, C.L.; Arnhold, S.; Jeong, J.J.; Park, J.H.; Eum, J.; Kim, B.; Fleckenstein, J.H. Monsoonal-type climate or land-use management: Understanding their role in the mobilization of nitrate and DOC in a mountainous catchment. *J. Hydrol.* **2013**, *507*, 149–162. [CrossRef]

33. Ha, K.J.; Park, S.K.; Kim, K.Y. On interannual characteristics of Climate Prediction Center merged analysis precipitation over the Korean peninsula during the summer monsoon season. *Int. J. Climatol.* **2005**, *25*, 99–116. [CrossRef]

34. Jung, I.W.; Bae, D.H.; Kim, G. Recent trends of mean and extreme precipitation in Korea. *Int. J. Climatol.* **2011**, *31*, 359–370. [CrossRef]

35. Seo, B.; Bogner, C.; Poppenborg, P.; Martin, E.; Hoffmeister, M.; Jun, M.; Koellner, T.; Reineking, B.; Shope, C.L.; Tenhunen, J. Deriving a per-field land use and land cover map in an agricultural mosaic catchment. *Earth Syst. Sci. Data* **2014**, *6*, 339–352. [CrossRef]

36. Jeon, M.S.; Kang, J.W. *Muddy Water Management and Agricultural Development Measures in the Watershed of Soyang Dam: Focused on Haean-myeon, Yanggu-gun*; Technical Report; Research Institute of Gangwon: Chuncheon, Korea, 2010.

37. Morgan, R.P.C.; Morgan, D.D.V.; Finney, H.J. A predictive model for the assessment of soil erosion risk. *J. Agric. Eng. Res.* **1984**, *30*, 245–253. [CrossRef]

38. Morgan, R.P.C. A simple approach to soil loss prediction: A revised Morgan–Morgan–Finney model. *CATENA* **2001**, *44*, 305–322. [CrossRef]

39. Vigiak, O.; Okoba, B.O.; Sterk, G.; Groenenberg, S. Modelling catchment-scale erosion patterns in the East African Highlands. *Earth Surf. Process. Landf.* **2005**, *30*, 183–196. [CrossRef]

40. Morgan, R.P.C.; Duzant, J.H. Modified MMF (Morgan–Morgan–Finney) model for evaluating effects of crops and vegetation cover on soil erosion. *Earth Surf. Process. Landf.* **2008**, *32*, 90–106. [CrossRef]

41. Lilhare, R.; Garg, V.; Nikam, B. Application of GIS-Coupled Modified MMF Model to Estimate Sediment Yield on a Watershed Scale. *J. Hydrol. Eng.* **2014**, *20*, C5014002. [CrossRef]

42. Choi, K.; Huwe, B.; Reineking, B. Commentary on "Modified MMF (Morgan–Morgan–Finney) Model for Evaluating Effects of Crops and Vegetation Cover on Soil Erosion" by Morgan and Duzant (2008). *arXiv* **2016**, arXiv:1612.08899.

43. ORNL DAAC. *MODIS Collection 5 Land Products Global Subsetting and Visualization Tool*; ORNL DAAC: Oak Ridge, TN, USA. Available online: https://doi.org/10.3334/ORNLDAAC/1379 (accessed on 9 June 2017). [CrossRef]

44. Schaap, M.G.; Leij, F.J.; van Genuchten, M.T. ROSETTA: A computer program for estimating soil hydraulic parameters with hierarchical pedotransfer functions. *J. Hydrol.* **2001**, *251*, 163–176. [CrossRef]

45. ORNL DAAC. *MODIS Collection 6 Land Products Global Subsetting and Visualization Tool*; ORNL DAAC: Oak Ridge, TN, USA. Available online: https://doi.org/10.3334/ORNLDAAC/1557 (accessed on 14 November 2017). [CrossRef]

46. Didan, K. *MOD13Q1 MODIS/Terra Vegetation Indices 16-Day L3 Global 250m SIN Grid V006*; NASA EOSDIS Land Processes DAAC: Oak Ridge, TN, USA, 2015.

47. Gutman, G.; Ignatov, A. The derivation of the green vegetation fraction from NOAA/AVHRR data for use in numerical weather prediction models. *Int. J. Remote Sens.* **1998**, *19*, 1533–1543. [CrossRef]

48. Rural Development Administration of South Korea. Agric. Technol. Portal (Nongsaro). 2018. Available online: http://nongsaro.go.kr/portal/farmTechMain.ps?menuId=PS00002 (accessed on 31 July 2018).

49. Morgan, R.P.C. *Soil Erosion and Conservation*; Blackwell Publishing: Malden, MA, USA, 2005; pp. 45–66.

50. Sobol', I.M. Sensitivity Analysis for Nonlinear Mathematical Models. *Math. Model. Comput. Exp.* **1993**, *1*, 407–414.

51. Saltelli, A. Making best use of model evaluations to compute sensitivity indices. *Comput. Phys. Commun.* **2002**, *145*, 280–297. [CrossRef]

52. Saltelli, A.; Annoni, P.; Azzini, I.; Campolongo, F.; Ratto, M.; Tarantola, S. Variance based sensitivity analysis of model output. Design and estimator for the total sensitivity index. *Comput. Phys. Commun.* **2010**, *181*, 259–270. [CrossRef]

53. Nossent, J.; Elsen, P.; Bauwens, W. Sobol' sensitivity analysis of a complex environmental model. *Environ. Model. Softw.* **2011**, *26*, 1515–1525. [CrossRef]

54. Qi, W.; Zhang, C.; Chu, J.; Zhou, H. Sobol's sensitivity analysis for TOPMODEL hydrological model: A case study for the Biliu River Basin, China. *J. Hydrol. Environ. Res.* **2013**, *1*, 1–10.

55. Saltelli, A.; Annoni, P. How to avoid a perfunctory sensitivity analysis. *Environ. Model. Softw.* **2010**, *25*, 1508–1517. [CrossRef]

56. Brooks, E.S.; Boll, J.; McDaniel, P.A. A hillslope-scale experiment to measure lateral saturated hydraulic conductivity. *Water Resour. Res.* **2004**, *40*, W04208. [CrossRef]

57. Iooss, B.; Janon, A.; Pujol, G.; Boumhaout, K.; Veiga, S.D.; Delage, T.; Fruth, J.; Gilquin, L.; Guillaume, J.; Le Gratiet, L.; et al. Sensitivity: Global Sensitivity Analysis of Model Outputs; R package version 1.15.1; 2018. Available online: https://CRAN.R-project.org/package=sensitivity (accessed on 8 August 2018).

58. R Core Team. *R: A Language and Environment for Statistical Computing*; R Foundation for Statistical Computing: Vienna, Austria, 2018.

59. Storn, R.; Price, K. Differential Evolution—A Simple and Efficient Heuristic for global Optimization over Continuous Spaces. *J. Glob. Optim.* **1997**, *11*, 341–359. [CrossRef]

60. Price, K.V.; Storn, R.M.; Lampinen, J.A. *Differential Evolution—A Practical Approach to Global Optimization*; Springer: Berlin/Heidelberg, Germany, 2006.

61. Ardia, D.; Mullen, K.M.; Peterson, B.G.; Ulrich, J. 'DEoptim': Differential Evolution in 'R', version 2.2-4. 2016. Available online: https://cran.r-project.org/web/packages/DEoptim (accessed on 5 May 2019).

62. Joseph, J.F.; Guillaume, J.H.A. Using a parallelized MCMC algorithm in R to identify appropriate likelihood functions for SWAT. *Environ. Model. Softw.* **2013**, *46*, 292–298. [CrossRef]

63. Zheng, F.; Zecchin, A.C.; Simpson, A.R. Investigating the run-time searching behavior of the differential evolution algorithm applied to water distribution system optimization. *Environ. Model. Softw.* **2015**, *69*, 292–307. [CrossRef]

64. Mullen, K.; Ardia, D.; Gil, D.; Windover, D.; Cline, J. DEoptim: An R Package for Global Optimization by Differential Evolution. *J. Stat. Softw.* **2011**, *40*, 1–26. [CrossRef]

65. Nash, J.E.; Sutcliffe, J.V. River flow forecasting through conceptual models part I—A discussion of principles. *J. Hydrol.* **1970**, *10*, 282–290. [CrossRef]

66. Moriasi, D.N.; Arnold, J.G.; Van Liew, M.W.; Bingner, R.L.; Harmel, R.D.; Veith, T.L. Model evaluation guidelines for systematic quantification of accuracy in watershed simulations. *Trans. ASABE* **2007**, *50*, 885–900. [CrossRef]

67. Moriasi, D.N.; Gitau, M.W.; Pai, N.; Daggupati, P. Hydrologic and Water Quality Models: Performance Measures and Evaluation Criteria. *Trans. ASABE* **2015**, *58*, 1763–1785. [CrossRef]

68. Mauricio Zambrano-Bigiarini. hydroGOF: Goodness-of-Fit Functions for Comparison of Simulated and Observed Hydrological Time Series; R Package Version 0.3-10. 2017. Available online: https://cran.r-project.org/web/packages/hydroGOF (accessed on 5 May 2019).

69. OECD. *Environmental Indicators for Agriculture: Methods and Results*; OECD Publishing: Paris, France, 2001; Volume 3.

70. OECD. *Environmental Performance of Agriculture in OECD Countries Since 1990*; OECD Publishing: Paris, France, 2008.

71. Tucker, G.E.; Whipple, K.X. Topographic outcomes predicted by stream erosion models: Sensitivity analysis and intermodel comparison. *J. Geophys. Res. Solid Earth* **2002**, *107*, 1–16. [CrossRef]

72. Neitsch, S.L.; Arnold, J.G.; Kiniry, J.R.; Williams, J.R. *Soil and Water Assessment Tool Theoretical Documentation Version 2009*; Technical Report; Texas Water Resources Institute: College Station, TX, USA, 2011.

73. Lee, J.Y. A Hydrological Analysis of Current Status of Turbid Water in Soyang River and Its Mitigation. *J. Soil Groundw. Environ.* **2008**, *13*, 85–92.

74. Kim, J.H.; Choi, H.T.; Lim, H.G. Analysis of Suspended Solid Generation with Rainfall-Runoff Events in a Small Forest Watershed. *J. Environ. Sci. Int.* **2015**, *24*, 1617–1627. [CrossRef]

75. Gellis, A.C. Factors influencing storm-generated suspended-sediment concentrations and loads in four basins of contrasting land use, humid-tropical Puerto Rico. *CATENA* **2013**, *104*, 39–57. [CrossRef]

76. Vercruysse, K.; Grabowski, R.C.; Rickson, R.J. Suspended sediment transport dynamics in rivers: Multi-scale drivers of temporal variation. *Earth Sci. Rev.* **2017**, *166*, 38–52. [CrossRef]

77. Lee, J.M.; Park, Y.S.; Kum, D.; Jung, Y.; Kim, B.; Hwang, S.J.; Kim, H.B.; Kim, C.; Lim, K.J. Assessing the effect of watershed slopes on recharge/baseflow and soil erosion. *Paddy Water Environ.* **2014**, *12*, 169–183. [CrossRef]

78. Lim, K.J.; Sagong, M.; Engel, B.A.; Tang, Z.; Choi, J.; Kim, K.S. GIS-based sediment assessment tool. *CATENA* **2005**, *64*, 61–80. [CrossRef]

79. Cooper, J.R.; Gilliam, J.W.; Daniels, R.B.; Robarge, W.P. Riparian Areas as Filters for Agricultural Sediment. *Soil Sci. Soc. Am. J.* **1987**, *51*, 416–420. [CrossRef]

80. Osborne, L.L.; Kovacic, D.A. Riparian vegetated buffer strips in water-quality restoration and stream management. *Freshwater Biol.* **1993**, *29*, 243–258. [CrossRef]

81. Tollner, E.W.; Barfield, B.J.; Haan, C.T.; Kao, T.Y. Suspended Sediment Filtration Capacity of Simulated Vegetation. *Trans. ASAE* **1976**, *19*, 678–682. [CrossRef]

82. Meyer, L.D.; Wischmeier, W.H. Mathematical simulation of the process of soil erosion by water. *Trans. ASAE* **1969**, *12*, 754–758. [CrossRef]

Article

Scaling-Up Conservation Agriculture Production System with Drip Irrigation by Integrating MCE Technique and the APEX Model

Tewodros Assefa [1,*], Manoj Jha [2], Abeyou W. Worqlul [3], Manuel Reyes [4] and Seifu Tilahun [1]

[1] Faculty of Civil and Water Resource Engineering, Bahir Dar University, Bahir Dar 26, Ethiopia; sat86@cornell.edu

[2] Department of Civil, Architectural and Environmental Engineering, North Carolina A&T State University, Greensboro, NC 27411, USA; mkjha@ncat.edu

[3] Texas A&M AgriLife Research, Temple, TX 76502, USA; aworqlul@brc.tamus.edu

[4] Sustainable Intensification Innovation Lab (SIIL), Kansas State University, Manhattan, KS 66506, USA; mannyreyes@ksu.edu

* Correspondence: tassefa@aggies.ncat.edu; Tel.: +25-191-210-0610

Received: 23 August 2019; Accepted: 18 September 2019; Published: 27 September 2019

Abstract: The conservation agriculture production system (CAPS) approach with drip irrigation has proven to have the potential to improve water management and food production in Ethiopia. A method of scaling-up crop yield under CAPS with drip irrigation is developed by integrating a biophysical model: APEX (agricultural policy environmental eXtender), and a Geographic Information System (GIS)-based multi-criteria evaluation (MCE) technique. Topography, land use, proximity to road networks, and population density were considered in identifying potentially irrigable land. Weather and soil texture data were used to delineate unique climate zones with similar soil properties for crop yield simulation using well-calibrated crop model parameters. Crops water demand for the cropping periods was used to determine groundwater potential for irrigation. The calibrated APEX crop model was then used to predict crop yield across the different climatic and soil zones. The MCE technique identified about 18.7 Mha of land (16.7% of the total landmass) as irrigable land in Ethiopia. Oromia has the highest irrigable land in the nation (35.4% of the irrigable land) when compared to other regional states. Groundwater could supply a significant amount of the irrigable land for dry season production under CAPS with drip irrigation for the various vegetables tested at the experimental sites with about 2.3 Mha, 3.5 Mha, 1.6 Mha, and 1.4 Mha of the irrigable land available to produce garlic, onion, cabbage, and tomato, respectively. When comparing regional states, Oromia had the highest groundwater potential (40.9% of total potential) followed by Amhara (20%) and Southern Nations, Nationalities, and Peoples (16%). CAPS with drip irrigation significantly increased groundwater potential for irrigation when compared to CTPS (conventional tillage production system) with traditional irrigation practice (i.e., 0.6 Mha under CTPS versus 2.2 Mha under CAPS on average). Similarly, CAPS with drip irrigation depicted significant improvement in crop productivity when compared to CTPS. APEX simulation of the average fresh vegetable yield on the irrigable land under CAPS with drip irrigation ranged from 1.8–2.8 t/ha, 1.4–2.2 t/ha, 5.5–15.7 t/ha, and 8.3–12.9 t/ha for garlic, onion, tomato, and cabbage, respectively. CAPS with drip irrigation technology could improve groundwater potential for irrigation up to five folds and intensify crop productivity by up to three to four folds across the nation.

Keywords: scaling-up conservation agriculture; drip irrigation; groundwater potential; sustainable intensification; Ethiopia

---

1. Introduction

Crop production in Ethiopia is constrained with several challenges that cause low productivity and economic growth in the region. Soil degradation in the form of soil erosion and decline of soil fertility is the major constraint for crop production [1]. The alarming rise in population caused the exploitation of the rainforest and grasslands in the region to increase cultivated lands, which resulted in soil degradation and deterioration of the environment [2–4]. Crop production in the nation is mainly a rainfed system using traditional farming practices [2]. The expansion of cultivated land at the expense of forest, bushes, and grassland is not a feasible option to sustain crop production let alone increased productivity. Instead, with the current poor soil and water management practices, it contributes to lower production efficiency [3].

On the other hand, rainfall variability is a great concern for a rainfed agricultural system in Ethiopia [5]. Customized local strategies are needed to maximize food supply and enhance the ecosystem at the same time. One such strategy is to enable dry season production to address the adverse effects of rainfall variability. The conservation agriculture production system (CAPS), which promotes no-till, mulching, and diverse cropping, has been shown to provide higher water use efficiency in addition to improving soil fertility [6–9]. Similarly, adoption efficient water application technologies can increase water use efficiency [10]. In relation to irrigation technology, drip irrigation is considered the most efficient water application technology [11,12]. CAPS combined with drip irrigation constitutes efficient soil and water management technology, which helps to maximize the potential of water resources and consequently increase productivity in the region [8]. Another concern is the lack of knowledge of potential to expand irrigated agriculture and maximize production. Worqlul et al., [3] indicated that less than 5% of the potentially irrigable lands are currently under irrigation.

While the positive impacts of CAPS with drip irrigation have been identified, expanding the impact to a large-scale adaptation on a country level and linking it with water resources availability would provide substantial and very useful information to policymakers in a decision-making process to improve the agriculture systems in the nation. Assessments of potentially irrigable land, corresponding crop productivity, and availability of water resources are essential components for the scale-up of CAPS with drip irrigation technology. There are few quantitative studies [3], that provided country-level irrigation potential assessment under the conventional tillage production system (CTPS) with traditional irrigation practice. However, no literature was found for a large-scale adaptation of CAPS with drip irrigation. This study attempts to examine the country-level adaptation of CAPS with drip irrigation technology for its impact on groundwater potential and crop productivity based on experimental results presented in Assefa et al. [8]. The specific objectives were to (1) assess potentially irrigable land using the multi-criteria evaluation (MCE) technique, (2) scale-up crop yields by integrating the MCE technique and a biophysical model field-scale prediction, and (3) asses groundwater irrigation potential for dry season production. The analyses were made for garlic, onion, tomato, and cabbage which are commonly grown vegetables in Ethiopia [13].

2. Materials and Methods

2.1. Study Area

This study was conducted in Ethiopia, the second-largest populated country in the entire continent of Africa, next to Nigeria (Figure 1). The landmass of the country is approximately 110 million ha, and elevation ranges from 160 m to 4530 m above mean sea level [3]. Climate variability (as it pertains to variability in rainfall and temperature) was observed to be very high in Ethiopia (i.e., 15% to 50% coefficient of variation for rainfall and 1.6°C annual average rise) based on the long-term (1955–2015) evaluation of climate data. This poses major risks to rainfed crop production [14,15] which is the dominant agriculture practice in Ethiopia [16]. The southwestern portion of the country receives about 2400 mm of rainfall, whereas northeastern and southeastern lowland receives less than 500 mm per year [17]. There are three seasons in the year locally known as Kiremt (main rainfall season), Belg (small rainfall season), and Bega (dry season) [18].

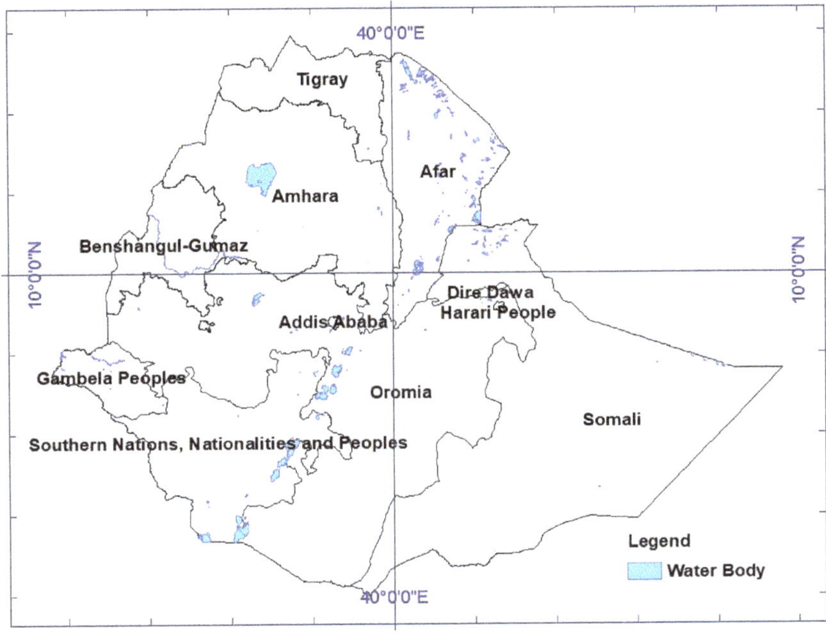

Figure 1. Ethiopia and its administrative regions with water bodies.

2.2. Scale-Up of Field-Scale Parameters

The MCE technique and agricultural policy environmental eXtender (APEX) model was used to scale-up the field-scale CAPS with drip irrigation (Figure 2). MCE was used to identify potentially irrigable lands in the country based on various factors that affect irrigated agriculture. MCE is an emerging approach that involves combining multiple variables to produce a single evaluation index for an intended purpose [2,19–26]. The MCE technique has been used for various applications including crop agriculture, water resource management, and other environmental studies [2,3,26–33].

A range of variables (Figure 2) were considered in this study to identify potentially irrigable land in the country. These factors include topography (slope), land use/cover, proximity to road networks, and population density. Topography affects the choice of irrigable land as it affects irrigation practices [3]. Digital elevation model (DEM) data with 30 m resolution, was used to derive the landscape slope for the entire nation. Land use/cover data also provide a vital figure in the selection of economically productive lands for irrigated agriculture. Similarly, population density and proximity to road networks were used to account market accessibility to support irrigated agriculture. Euclidean distance was calculated to establish the proximity criteria to road networks. Factors were reclassified into various suitability classes depending on the Food and Agricultural Organization [34] guidelines: Highly suitable (S1), moderately suitable (S2), marginally suitable (S3), currently unsuitable (S4), and permanently unsuitable or constraint (N1). The equal interval ranging technique was used to reclassify population density and proximity to road networks based on Worqlul et al., [3]. The pairwise method was used to compare each factor one-to-one and weights were scaled using works of Saaty [35] and Worqlul et al., [3]. The pairwise method is a relatively unbiased ranking technique [2,27] and applied to weigh each factor considered in this analysis. The pairwise technique, Saaty [35], makes use of a scale broken down from 1 to 9 indicating the equal and absolute importance of a factor when compared to a one-to-one basis, respectively. Consistency ratio was used to check the consistency of the pairwise matrix using Equation (1) described in Saaty [36]. The equal interval ranging technique was used to distribute the weight of each factor into the suitability classes. Factors were then combined using "weighted sum overlay" to produce a single evaluation index (0% to 100%) map, and constraints (permanently unsuitable lands) were excluded from the analysis. Combined weights of greater than 85% were considered to be potentially irrigable lands [2,3].

$$CR = CI/RI, \tag{1}$$

where CR = consistency ratio, CI = consistency index, and RI = consistency index of randomly generated matrix.

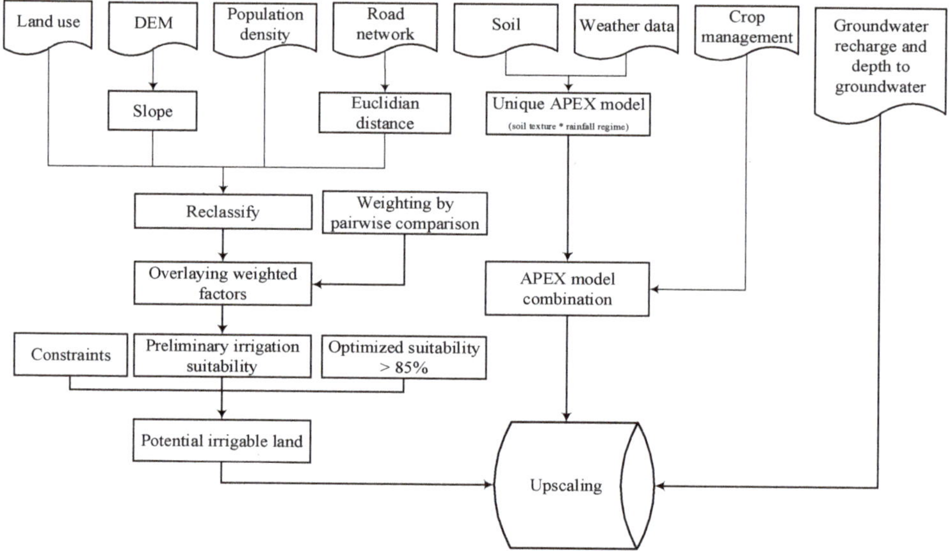

Figure 2. Method of scaling-up crop production under conservation agriculture production system (CAPS) with drip irrigation technology.

Soil and climate data were used to define unique areas for biophysical model development. Spatial variability of parameters and the effect of scaling needs to be carefully considered while upscaling modeling results [37–39]. Table 1 shows the type and source of data used to scale-up field-scale crop production to the national level. The mean annual rainfall point data from observed ground weather stations were used to compute spatial annual rainfall and classify the nation into different rainfall regimes. Similarly, soil texture data was used to classify the nation into various soil classes. Climate regimes and soil texture data were combined using the intersection of the ArcGIS overlay function to identify geographical equivalence zones of similar climate and soil (i.e., 39 zones) where each zone has the same soil texture and rainfall regimes. The APEX [40], a biophysical model, was set up on the unique climate and soil combinations to simulate crop yield. Weather data (rainfall, maximum and minimum temperature, wind speed, relative humidity, and solar radiation), soil characteristics, as well as vegetation and management practices are the main inputs to the APEX model [41]. The model is capable of evaluating the effects of soil and water conservation practices on hydrology, crop yield, and other environmental variables such as sediment, nutrient load, and soil organic carbon [42–44]. Proper calibration and validation of model parameters are essential steps for reliable predictions [45]. The APEX model was calibrated and validated for a few sites in Ethiopia using adequate field data from 13 experimental plots. The calibration results (i.e., model performances) are presented in Table 5 and

Table 9 of Assefa et al., [9] for hydrology and crop yield, respectively. Based on efficiency measures suggested in Moriasi et al., [46] and Wang et al. [47], the model performance was found within the range of acceptable to very good. Satisfactory model performance during calibration and validation provides greater confidence in the modeling results when evaluating various plausible scenarios for modeling prediction. The present study was built on the same model for its application in up-scaling the impact. Input data were changed based on the unique climate-soil combinations across Ethiopia, but the same model parameters were used which was established during the calibration. Heat unit scheduling (OPV7 = 1) was used with shortening cropping period compared to the experimental plots to capture crop growth variability across the unique regions.

Table 1. Data and sources for upscaling crop production to the county level.

Data	Source	Spatial Resolution (m)
Land use	World land use database (LADA), Food and Agricultural Organization (FAO), 2010	10,000
Soil	Africa Soil Information Service (AFSIS), 2015	250
Digital Elevation Model (DEM)	Unites States Geographical Survey (USGS), 2000 (2015 release)	30
Population density	Global gridded pupation database, 2000	1000
MODIS potential evapotranspiration (mm)	MOD16 Global Terrestrial Evapotranspiration data set (2000–2010)	1000
Potential borehole yield (L/s)	British Geological Survey (BGS), 2012	5000
Groundwater depth (m)	British Geological Survey (BGS), 2012	5000
Rainfall (mm)	Ethiopian National Meteorological Agency (ENMA), 2000 to 2010	-

Irrigation requirement of crops is mainly a function of reference evapotranspiration and rainfall, which are variable in space and time. Therefore, variable irrigation water volumes were applied in the model for each climate zone depending on the type of vegetables grown and weather conditions. The net irrigation requirement (NIR) for each of the vegetables was calculated depending on reference evapotranspiration (ETo), crop coefficients of each vegetable at mid-stage (Kc), irrigation application inefficiency, and effective rainfall amount (ER). Crop coefficients at the mid-stage of crop growth (Kc-mid) were obtained from Allen et al., [48] of the Food and Agricultural Organization (FAO) for various vegetables. The net irrigation requirement equation derived by Worqlul et al., [3] for the country using conventional irrigation inefficacy in Equation (2) was modified in this study (Equation (3)) to account for drip irrigation inefficiency. Howell [49] indicated that 95% efficiency can be attainable whereas 90% is the average efficiency for drip irrigation. Thus, 10% application inefficiency was considered for irrigation and some minor losses such as leaching [50]. Assefa et al. [8] showed that significant ($p \leq 0.05$) reduction of irrigation volume was observed in the conservation of agriculture (CA) experimental sites when compared to conventional tillage (CT) practice. Therefore, Equation (3) was further modified using linear coefficient (Cf) to account for the reduction of irrigation volume in CA practice by comparing the irrigation data from the experimental sites for CA and CT managements (Equation (4)) (net irrigation requirement for conservation practice, NIRc). The contribution of rainfall to soil moisture, effective rainfall (ER), in the growing season of vegetables was estimated using the United States Department of Agriculture Soil Conservation Service (USDA-SCS) method [51], which is a function of precipitation (P), see Equation (5a) and Equation (5b).

$$NIR = 1.6 \times Kc \times ETo - ER \tag{2}$$

$$NIR = 1.1 \times Kc \times ETo - ER \tag{3}$$

$$NIRc = 1.1 \times Cf \times Kc \times ETo - ER \tag{4}$$

$$ER = P \times \frac{125 - 0.2 \times P}{125}; \text{ For } P \leq 250 \text{ mm/m} \tag{5a}$$

$$ER = 125 + 0.1 \times P; \text{ For } P > 250 \text{ mm/m} \tag{5b}$$

where NIR, NIRc, Cf, Kc, ETo, P, and ER are the net irrigation requirement for the tilled system, net irrigation requirement for conservation agriculture, the coefficient of conservation agriculture, crop coefficient, reference evapotranspiration, precipitation, and effective rainfall, respectively.

Soil properties, weather data, net irrigation requirements, and cropping details were supplied to the well-calibrated APEX model in each unique zone. Then, crop yield simulation was integrated with the irrigable land to limit crop yield estimation only on the potentially irrigable land. Groundwater source of irrigation with a depth less than 30 m from the surface was considered in this study. The potential borehole yields and the potential numbers of wells that could be installed were used to estimate groundwater availability in the regions. Maintaining a one-kilometer clear distance between wells (i.e., the radius of influence) is suggested by Howsam and Carter [52] to estimate the potential numbers of wells that could be installed. Maintaining the radius of influence helps to avoid the groundwater drawdown effect of one well on another. The net irrigation requirement for CA practice, groundwater availability, and depth to groundwater were considered to determine the potential of groundwater wells in unique zones. Vegetable yields on the irrigable land were further constrained based on groundwater availability to identify the potential scale-up areas for CAPS with drip irrigation technology.

3. Results and Discussion

The results of scaling-up crop yield under CAPS with drip irrigation technology to country-level were presented into three categories: (1) Assessment of potentially irrigable land in the country using the MCE technique, (2) simulation of potential crop production under CAPS with drip irrigation using a well-calibrated APEX model, and (3) assessment of groundwater potential for dry season crop production.

3.1. Potentially Irrigable Land

Four basic factors (topography, land use, proximity to road networks, and population density) were considered in the MCE technique to identify potentially irrigable land in the nation. Topography in the nation ranges from 0% (flat land) to greater than 100% (steepest land which is about 0.07% of the landmass) (Figure 3a). The slope was reclassified into five categories based on Worqlul et al., [27]: Highly suitable (0%–2%), moderately suitable (2%–8%), marginally suitable (8%–12%), less suitable (12%–30%), and unsuitable (above 30%). The various land use classes in the nation (Figure 3b) were reclassified into four suitability classes based on Assefa et al. [2], Worqlul et al., [3], and FAO [53]: Highly suitable (agricultural land), moderately suitable (grassland), marginally suitable class (shrubs, bare land), and unsuitable class (forest, urban lands, wetlands, and water). Population density ranges from 0 to 69,350 persons per square kilometer (Figure 3c), whereas proximity to road network ranges from 0 to 118 km (Figure 3d).

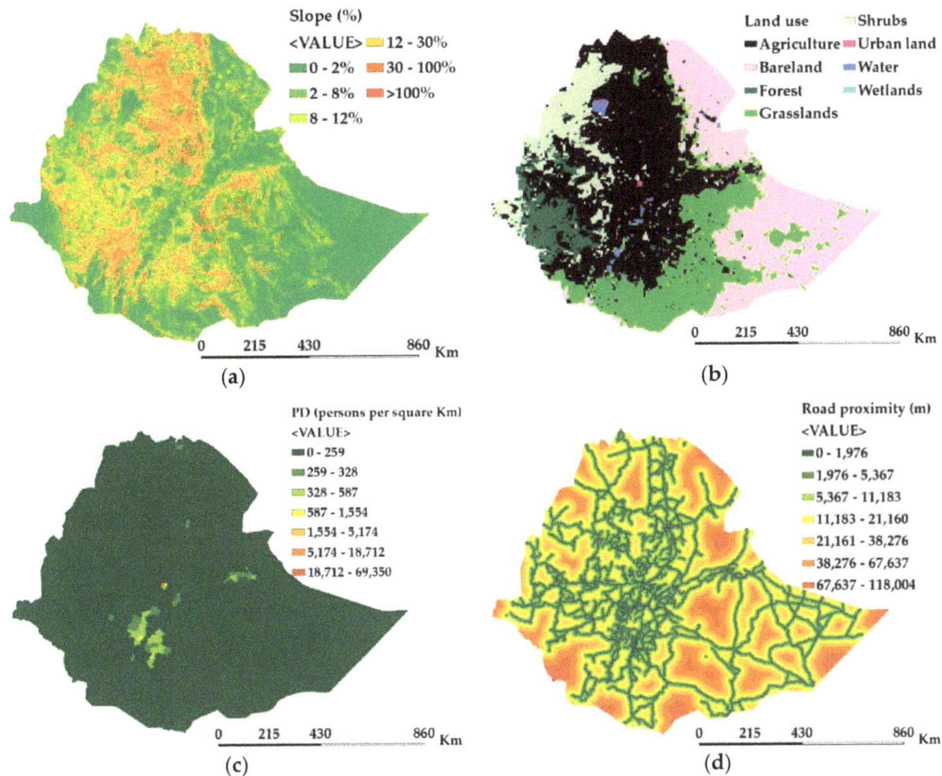

Figure 3. Irrigation suitability factors map: (**a**) Topography in terms of landscape slope, (**b**) land use, (**c**) population density (PD), and (**d**) distance to road networks.

The Eigenvector was computed as the nth root of individual factors' weight and then normalized with the cumulative Eigenvector to derive the final weights of factors (Table 2). Topography was found to be relatively the most influential factor in irrigated agriculture, which was consistent with Worqlul et al.,'s [3] result. Proximity to road networks and land use were found to be the second and third most influential factors in determining potentially irrigable land in the nation. The consistency ratio was found to be trustworthy (CR = 0.03 ≤ 0.2) based on Chen et al., [54] and Koczkodaj et al., [55]. The final weights of factors were distributed to the various suitability classes and factors were combined using a weighted sum overlay. An 85% threshold was used to obtain potentially irrigable land (Figure 4). About 18.7 Mha of land, 16.7% of the total landmass, was found to be potentially irrigable in the nation without considering soil and weather. The suitability ranges in Figure 4 cover different portions of the irrigable land: 85%–88% (76% of the irrigable land), 88%–91% (11% of the irrigable land), 91%–94% (1.5 of the irrigable land), 94%–97% (11.5% of the irrigable land), and 97%–100% (0.4% of the irrigable land).

Table 2. Pairwise matrix for calculation of the weight of factors.

Factors	Slope	Road Proximity	Population Density	Land Use	Eigenvector	Weight (%)
Slope	1.0	2.0	4.0	3.0	2.2	46.3
Road	1/2	1.0	3.0	2.0	1.3	27.5
Population density	1/4	1/3	1.0	1/3	0.4	8.5
Land use	1/3	1/2	3.0	1.0	0.8	17.6

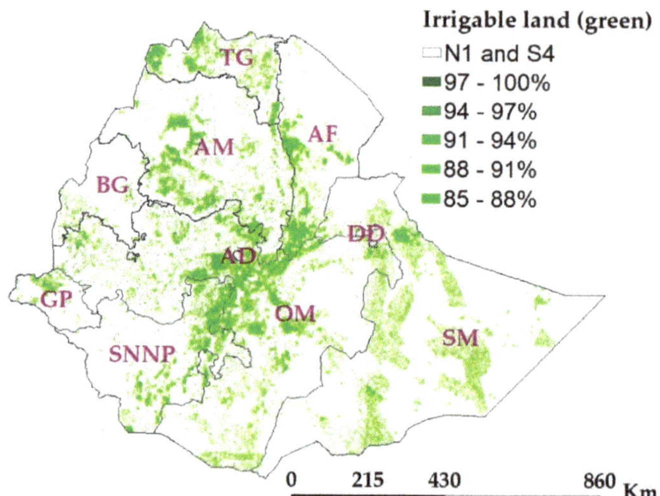

Figure 4. Potentially irrigable land (all green, weight ≥ 85); S4 (currently unsuitable areas)- weight < 85%; N1—constraint (permanently unsuitable areas); abbreviations in the map are administrative regions (TG—Tigray, AM—Amhara, AF—Afar, BG—Benshangul Gumaz, AD—Addis Ababa, DD—Dire Dawa, GP—Gambela Peoples, SNNP—Southern Nations, Nationalities and Peoples, and SM—Somali).

Irrigation demand of each vegetable was computed by considering the conservation agriculture principles, drip irrigation technology, water use of different vegetables, and weather conditions. Oromia regional state has the highest irrigable land (35.4%) when compared with other states. Figure 5 illustrates the degree of irrigation suitability (i.e., marginal, satisfactory, medium, high, and very high) for potentially irrigable lands: 18.7 Mha, 4.5 Mha, 2.5 Mha, 2.2 Mha, and 0.082 Mha at 85%, 88%, 91%, 94%, and 97% suitability classes, respectively.

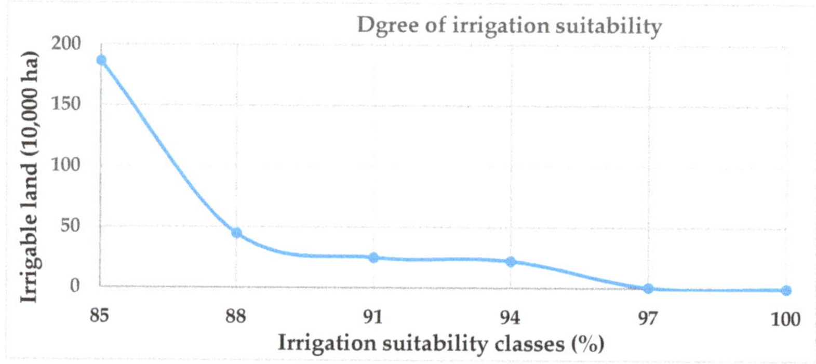

Figure 5. Degree of irrigation suitability for the potentially irrigable land.

3.2. Potential Crop Production under Conservation Agriculture

Figure 6a shows the various soil texture classes in the nation. The mean annual rainfall was computed spatially using inverse distance weighting interpolation from weather stations point data (Figure 6b), and the spatial rainfall was reclassified using natural breaks into eight rainfall zones (Figure 6c). Soil textures and rainfall zones were combined, which resulted in 39 unique regions for further analyses of crop yields. The APEX model was developed for each of 39 unique zones, which was defined using the soil texture classes and climate zones. Results were then aggregated as per administrative boundaries of Ethiopia (11 regional states) to provide input for decisionmakers in developing policy and implementation strategies for water resource and agriculture-related projects.

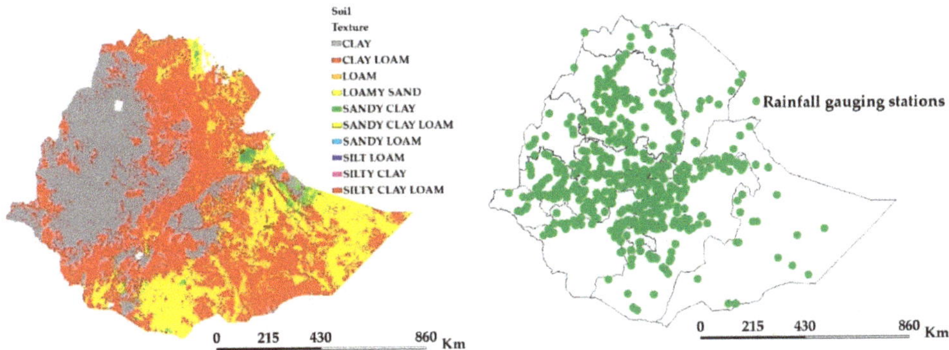

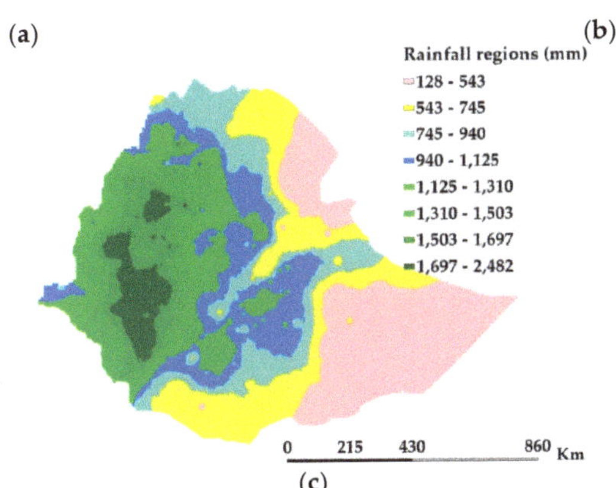

Figure 6. Soil textures (**a**), locations of rainfall gauging stations (**b**), and rainfall regions regimes (**c**).

Crop coefficients at the mid-stage of crop growth (Kc-mid) were obtained from Allen et al., [48] of the Food and Agricultural Organization (FAO) for various vegetables. These data indicate that more irrigation is needed for tomato during the mid-stage of growth followed by cabbage when compared to garlic and onion. Moderate resolution image spectrum (MODIS) potential evapotranspiration data (2000–2010) was used to estimate the net irrigation demand in the region during the dry season. The growing period used for garlic, onion, and cabbage was December through February whereas December through March was the growing season for tomato.

This study used these water use data for the various vegetables under CA and CT practices from Assefa et al., [8] and developed a linear irrigation coefficient, Cf, for each vegetable to account for irrigation volume reduction under CA during the calculation for the irrigation requirement. The value of Cf obtained from CA and CT comparison was 0.58, 0.54, 0.80, and 0.81 for garlic, onion, tomato, and cabbage, respectively. These coefficients explain the advantage of conservation practices over conventional tillage systems for irrigation water savings mainly due to mulch cover and no-till practice in CAPS plots minimized water loss through evaporation and runoff. Additionally, the water-saving in garlic and onions was higher when compared with cabbage and tomato. This could be due to the less leaf area of garlic and onion, which made the impact of CAPS significant in reducing water loss when compared with tomato and cabbage.

Net irrigation demand was computed for each vegetable over the irrigable land considering drip irrigation efficiency, crop coefficient, effective rainfall during the growing period, and irrigation coefficient. These data along with other inputs such as soils, weather data, cropping details, irrigation application rate, and crop water demand were supplied to the calibrated APEX model to estimate crop yield. Crop yield results were averaged for the simulation period (2000–2010) and limited to potentially irrigable land in the nation. The average fresh vegetable yield under CAPS ranged from 1.8–2.8 t ha^{-1} for garlic (Figure 7a), 1.4–2.2 t ha^{-1} for onions (Figure 7b), 5.5–15.7 t ha^{-1} for tomato (Figure 7c), and 8.3–12.9 t ha^{-1} for cabbage (Figure 7d). Crop productivity was found to be higher in Oromia and Amhara regions due to the combined effects of the weather and soil condition. The variation of yields for tomato was found relatively high when compared to other vegetables, possibly due to weather variations and the fact that tomato is more sensitive to cold weather. The maximum and minimum allowable temperature for tomato is 27°C and 10°C, respectively, for optimal crop growth.

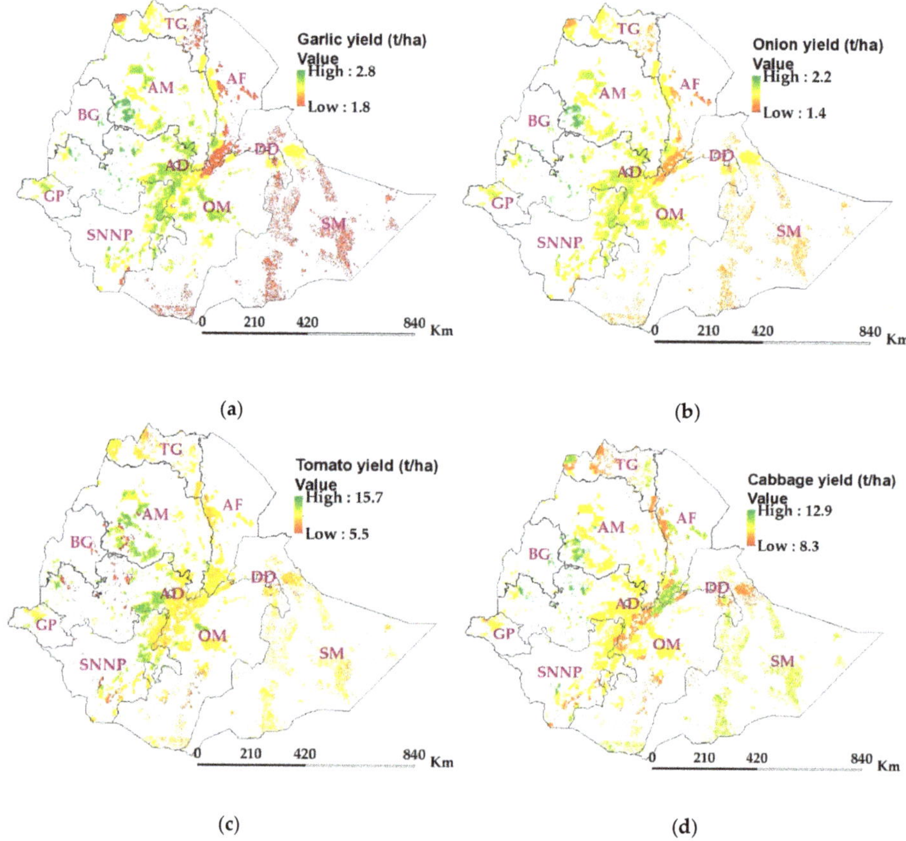

(a) (b)

(c) (d)

Figure 7. Potential crop yields over the irrigable lands (**a**) garlic, (**b**) onion, (**c**) tomato, and (**d**) cabbage. Abbreviations in the map are administrative regions (TG—Tigray, AM—Amhara, AF—Afar, BG—Benshangul Gumaz, AD—Addis Ababa, DD—Dire Dawa, GP—Gambela Peoples, SNNP—Southern Nations, Nationalities and Peoples, and SM—Somali).

3.3. Groundwater Potential for Crop Production under CAPS with Drip Irrigation

Groundwater depth of less than 30 m is considered feasible for irrigation in the nation Gebregziabher [56]. Thus, depth to groundwater less than 30 m were considered in this study for the estimation of groundwater potential. Worqlul et al., [3] validated the British Geological Survey (BGS) groundwater borehole yield estimates in the central part of Ethiopia using actual groundwater recharge data from the Agricultural Transformation Agency (ATA). The net irrigation requirements for crops were deducted from groundwater potential to identify areas where groundwater fully supports to produce vegetables during the dry season. Figure 8 depicts areas where groundwater potential can support to produce garlic, onion, tomato, and cabbage, respectively.

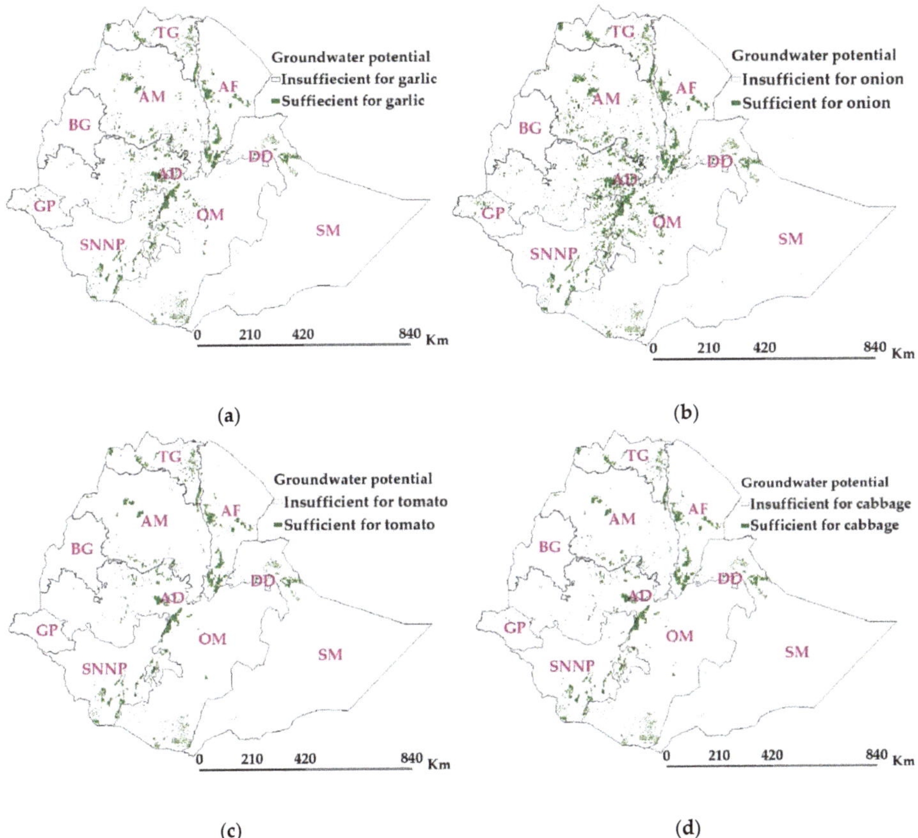

Figure 8. Crop yields over the irrigable lands (**a**) garlic, (**b**) onion, (**c**) tomato, and (**d**) cabbage. Abbreviations in the map are administrative regions (TG—Tigray, AM—Amhara, AF—Afar, BG—Benshangul Gumaz, AD—Addis Ababa, DD—Dire Dawa, GP—Gambela Peoples, SNNP—Southern Nations, Nationalities and Peoples, and SM—Somali).

Table 3 presents the potential of groundwater for different vegetables as a percentage of potentially irrigable land over the administrative regions. For instance, considering the Oromia region, groundwater is enough to irrigate 0.95 Mha, 1.5 Mha, 0.6 Mha, or 0.5 Mha if planting garlic, onion, tomato, or cabbage, respectively, from the potentially irrigable land (6.6 Mha) if planting garlic or onion. That means, 8.9% to 14.3% of the potential land in Oromia could be irrigated depending on the type of crop using groundwater if CAPS with drip irrigation is used. Similarly, 8.8% to 30% of the potential land in Amhara and 11.6% to 29.8% of the potential land in Southern Nations, Nationalities and Peoples (SNNP) could be irrigated using groundwater. Oromia has the highest groundwater potential (40.9% of total potential) followed by Amhara (20% of total potential), and SNNP (16%). At country level (aggregated from administrative regions), groundwater potential was found to support about 2.3 Mha (Figure 8a), 3.5 Mha (Figure 8b), 1.6 Mha (Figure 8c), and 1.4 Mha (Figure 8d) of land to produce garlic, onion, cabbage, and tomato, respectively in the dry season. Onion has relatively the least irrigation demand and thus has the highest production area coverage using groundwater, followed by garlic, whereas tomato and cabbage have relatively high irrigation demands and thus less area coverage for production using groundwater source.

Table 3. Irrigable land and potential of groundwater for various vegetables.

Administrative Region	Irrigable land (1000 ha)	Groundwater Potential on the Irrigable Land (1000 ha)			
		Garlic	Onion	Tomato	Cabbage
Addis Ababa	9.3	1.2	3.0	0.0	0.0
Afar	1539.0	255.5	303.6	239.8	236.0
Amhara	2628.0	459.2	787.3	291.4	230.0
Benshangul-Gumaz	327.0	9.6	20.0	5.9	2.7
Dire Dawa	16.7	1.5	1.5	1.5	1.5
Gambela Peoples	320.0	32.9	86.2	10.8	0.9
Harari People	11.4	0.0	0.0	0.0	0.0
Oromia	6621.0	946.7	1473.5	644.8	553.0
Somali	3990.0	49.0	55.2	48.2	46.8
SNNP	1910.0	369.7	570.0	254.9	222.0
Tigray	1326.0	152.0	175.3	148.0	146.5

Note: SNNP—Southern Nations, Nationalities and Peoples.

4. Conclusions

This study is the first of its kind in providing insight into the impacts of the large-scale adaptation of CAPS with drip irrigation on groundwater potential and crop productivity for common vegetables grown in Ethiopia. The results from the MCE technique indicated that there was substantial amount of land for irrigation using groundwater source (~17% of the total landmass). A comparison between suitable areas for irrigation and groundwater potential showed that a modest amount of land (up to 19% of the irrigable land) could be irrigated under CAPS and drip irrigation. The potential of groundwater, however, is a limiting factor to expand irrigated agriculture on suitable lands. Oromia and Amhara regional states provided about 61% of the nation's groundwater potential for irrigation, hence it would be a wise choice for policymakers to consider these results in expanding irrigated agriculture for dry season crop productions.

A comparison between groundwater potential results under CAPS with drip irrigation (1.4 to 3.5 Mha) and CTPS [3], showed that CAPS with drip irrigation significantly increased groundwater potential for irrigation (i.e., 0.6 Mha under CTPS versus 2.2 Mha under CAPS on average). Groundwater potential could be further improved if irrigation scheduling was incorporated with the drip application system. Garlic and onion could be produced in relatively larger areas compared to tomatoes and cabbages due to relatively lower irrigation demand. In addition, CAPS with drip irrigation could significantly improve crop productivity in the nation when compared to CTPS with traditional irrigation. Production potential under CAPS with drip irrigation for cabbage (8.3 t ha^{-1} to 12.9 t ha^{-1}) was substantially higher than CTPS, [57], which is 7.9 t ha^{-1} for the national average. Therefore, CAPS with drip irrigation is a feasible strategy to improve groundwater potential and crop productivity in the nation. Hence, policymakers should consider CAPS with drip irrigation in expanding small-scale irrigated agriculture.

Author Contributions: T.A. contributed to the conceptual design, data collection and acquisition, data analysis, and writing the manuscript. M.J. contributed to the conceptual design, data acquisition, data analysis, and revising the manuscript for scientific content. A.W.W. contributed to the conceptual design, data acquisition, data analysis, and revising the manuscript for scientific content. M.R. contributed to the conceptual design, data acquisition, and analysis. S.T. contributed to the conceptual design, data acquisition, data analysis, and revising the manuscript for scientific content.

Funding: This research and publication are made possible by the generous support of the American people through support by the United States Agency for International Development Feed the Future Innovation Labs for Collaborative Research on Small Scale Irrigation (Cooperative Agreement No. AID-OAA-A-13-0005, Texas A&M University) and Sustainable Intensification (Cooperative Agreement No. AID-OAA-L-14-00006, Kansas State University). The opinions expressed herein are those of the author(s) and do not necessarily reflect the views of the U.S. Agency for International Development.

Acknowledgments: We would like to acknowledge the International Water Management Institute (IWMI), and Ethiopian National Meteorological Agency (ENMA) for providing quality data for this research.

Conflicts of Interest: The authors declare no conflict of interest.

References

1. Tesfa, A.; Mekuriaw, S. The effect of land degradation on farm size dynamics and crop-livestock farming system in Ethiopia: A Review. *Open J. Soil Sci.* **2014**, *4*, 1. [CrossRef]

2. Assefa, T.; Jha, M.; Reyes, M.; Srinivasan, R.; Worqlul, A.W. Assessment of Suitable Areas for Home Gardens for Irrigation Potential, Water Availability, and Water-Lifting Technologies. *Water* **2018**, *10*, 495. [CrossRef]

3. Worqlul, A.W.; Jeong, J.; Dile, Y.T.; Osorio, J.; Schmitter, P.; Gerik, T.; Srinivasan, R.; Clark, N. Assessing potential land suitable for surface irrigation using groundwater in Ethiopia. *Appl. Geogr.* **2017**, *85*, 1–13. [CrossRef]

4. Getahun, K.; Van Rompaey, A.; Van Turnhout, P.; Poesen, J. Factors controlling patterns of deforestation in moist evergreen Afromontane forests of Southwest Ethiopia. *For. Ecol. Manag.* **2013**, *304*, 171–181. [CrossRef]

5. Bekele, A.E. Five key constraints to small scale irrigation development in Ethiopia: Socio-Economic View. *Glob. Adv. Res. J.* **2014**, *3*, 441–444.

6. Giller, K.E.; Witter, E.; Corbeels, M.; Tittonell, P. Conservation agriculture and smallholder farming in Africa: The heretics' view. *Field Crop. Res.* **2009**, *114*, 23–34. [CrossRef]

7. Assefa, T.T.; Jha, M.K.; Reyes, M.R.; Schimmel, K.; Tilahun, S.A. Commercial Home Gardens under Conservation Agriculture and Drip Irrigation for Small Holder Farming in sub-Saharan Africa. In Proceedings of the 2017 ASABE Annual International Meeting, Spokane, WC, USA, 16–19 July 2017; p. 1.

8. Assefa, T.; Jha, M.; Reyes, M.; Tilahun, S.; Worqlul, A.W. Experimental Evaluation of Conservation Agriculture with Drip Irrigation for Water Productivity in Sub-Saharan Africa. *Water* **2019**, *11*, 530. [CrossRef]

9. Assefa, T.; Jha, M.; Reyes, M.; Worqlul, A. Modeling the Impacts of Conservation Agriculture with a Drip Irrigation System on the Hydrology and Water Management in Sub-Saharan Africa. *Sustainability* **2018**, *10*, 4763. [CrossRef]

10. Ward, F.A.; Pulido-Velazquez, M. Water conservation in irrigation can increase water use. *Proc. Natl. Acad. Sci.* **2008**, *105*, 18215–18220. [CrossRef]

11. Megersa, G.; Abdulahi, J. Irrigation system in Israel: A review. *Int. J. Water Resour. Environ. Eng.* **2015**, *7*, 29–37.

12. Assefa, T.T. *Experimental and Modeling Evaluation of Conservation Agriculture with Drip Irrigation for Small-Scale Agriculture in Sub-Saharan Africa*; North Carolina Agricultural and Technical State University: Greensboro, NC, USA, 2018.

13. Emana, B.; Afari-Sefa, V.; Dinssa, F.F.; Ayana, A.; Balemi, T.; Temesgen, M. Characterization and assessment of vegetable production and marketing systems in the Humid Tropics of Ethiopia. *Q. J. Int. Agric.* **2015**, *54*, 163–187.

14. Abebe, G. Long-term climate data description in Ethiopia. *Data Brief* **2017**, *14*, 371–392. [CrossRef] [PubMed]

15. Kassie, B.; Rötter, R.; Hengsdijk, H.; Asseng, S.; Van Ittersum, M.; Kahiluoto, H.; Van Keulen, H. Climate variability and change in the Central Rift Valley of Ethiopia: Challenges for rainfed crop production. *J. Agric. Sci.* **2014**, *152*, 58–74. [CrossRef]

16. Arndt, C.; Robinson, S.; Willenbockel, D. Ethiopia's growth prospects in a changing climate: A stochastic general equilibrium approach. *Glob. Environ. Chang.* **2011**, *21*, 701–710. [CrossRef]

17. Fekadu, K. Ethiopian seasonal rainfall variability and prediction using canonical correlation analysis (CCA). *Earth Sci* **2015**, *4*, 112–119. [CrossRef]

18. Degefu, W. *Some Aspects of Meteorological Drought in Ethiopia*; Drought and Hunger in Africa; Cambridge University Press: Cambridge, UK, 1987.

19. Ayalew, G. Land suitability evaluation for surface and sprinkler irrigation using Geographical Information System (GIS) in Guang Watershed, Highlands of Ethiopia. *J. Environ. Earth Sci.* **2014**, *4*, 140–149.

20. Malczewski, J. GIS-based multicriteria decision analysis: A survey of the literature. *Int. J. Geogr. Inf. Sci.* **2006**, *20*, 703–726. [CrossRef]

21. Laaribi, A.; Chevallier, J.; Martel, J.M. A spatial decision aid: A multicriterion evaluation approach. *Comput. Environ. Urban Syst.* **1996**, *20*, 351–366. [CrossRef]

22. Malczewksi, J.; Ogryczak, W. The multiple criteria location problem: 2. Preference-based techniques and interactive decision support. *Environ. Plan. A* **1996**, *28*, 69–98. [CrossRef]

23. Chakhar, S.; Martel, J.M. Enhancing geographical information systems capabilities with multi-criteria evaluation functions. *J. Geogr. Inf. Decis. Anal.* **2003**, *7*, 47–71.

24. Carver, S.J. Integrating multi-criteria evaluation with geographical information systems. *Int. J. Geogr. Inf. Syst.* **1991**, *5*, 321–339. [CrossRef]

25. Chuvieco, E. Integration of linear programming and GIS for land-use modelling. *Int. J. Geogr. Inf. Sci.* **1993**, *7*, 71–83. [CrossRef]

26. Assefa, T.T.; Jha, M.K.; Tilahun, S.A.; Yetbarek, E.; Adem, A.A.; Wale, A. Identification of erosion hotspot area using GIS and MCE technique for koga watershed in the upper blue Nile Basin, Ethiopia. *Am. J. Environ. Sci.* **2015**, *11*, 245–255. [CrossRef]

27. Worqlul, A.W.; Collick, A.S.; Rossiter, D.G.; Langan, S.; Steenhuis, T.S. Assessment of surface water irrigation potential in the Ethiopian highlands: The Lake Tana Basin. *Catena* **2015**, *129*, 76–85. [CrossRef]

28. Teka, K.; Van Rompaey, A.; Poesen, J. Land suitability assessment for different irrigation methods in Korir Watershed, Northern Ethiopia. *J. Drylands* **2010**, *3*, 214–219.

29. Maddahi, Z.; Jalalian, A.; Zarkesh, M.M.K.; Honarjo, N. Land suitability analysis for rice cultivation using multi criteria evaluation approach and GIS. *Eur. J. Exp. Biol.* **2014**, *4*, 639–648.

30. Baniya, M.S.N. Land Suitability Evaluation Using GIS for Vegetable Crops in Kathmandu Valley/Nepal. Ph.D. Thesis, Institute of Horticulture Science, Humboldt-Universität zu Berlin, Berlin, Germany, 2008.

31. Hossain, M.S.; Chowdhury, S.R.; Das, N.G.; Rahaman, M.M. Multi-criteria evaluation approach to GIS-based land-suitability classification for tilapia farming in Bangladesh. *Aquac. Int.* **2007**, *15*, 425–443. [CrossRef]

32. Chen, Y.; Khan, S.; Padar, Z. Irrigation intensification or extensification assessment: A GIS-based spatial fuzzy multi-criteria evaluation. In Proceedings of the 8th International Symposium on Spatial Accuracy Assessment in Natural Resources and Environmental Sciences, Shanghai, China, 25–27 June 2018; pp. 309–318.

33. Tu, Q.; Li, H.; Wang, X.; Chen, C.; Luo, Y.; Dwomoh, F.A. Multi-criteria evaluation of small-scale sprinkler irrigation systems using Grey relational analysis. *Water Resour. Manag.* **2014**, *28*, 4665–4684. [CrossRef]

34. FAO. *A Framework for Land Evaluation*; Soils bulletin No. 32; FAO: Rome, Italy, 1976.

35. Saaty, T.L. A scaling method for priorities in hierarchical structures. *J. Math. Psychol.* **1977**, *15*, 234–281. [CrossRef]

36. Saaty, R.W. The analytic hierarchy process—What it is and how it is used. *Math. Model.* **1987**, *9*, 161–176. [CrossRef]

37. Baffaut, C.; Dabney, S.M.; Smolen, M.D.; Youssef, M.A.; Bonta, J.V.; Chu, M.L.; Guzman, J.A.; Shedekar, V.S.; Jha, M.K.; Arnold, J.G. Hydrologic and water quality modeling: Spatial and temporal considerations. *Trans. ASABE* **2015**, *58*, 1661–1680.

38. Guzman, J.A.; Shirmohammadi, A.; Sadeghi, A.M.; Wang, X.; Chu, M.L.; Jha, M.K.; Parajuli, P.B.; Harmel, R.D.; Khare, Y.P.; Hernandez, J.E. Uncertainty considerations in calibration and validation of hydrologic and water quality models. *Trans. ASABE* **2015**, *58*, 1745–1762.

39. Jha, M.K.; Gassman, P.W.; Secchi, S.; Gu, R.; Arnold, J.G. Impact of Watershed Subdivision Level on Flows, Sediment Loads, and Nutrient Losses Predicted by SWAT. *J. Am. Water Resour. Assoc.* **2004**, *40*, 811–825. [CrossRef]

40. Williams, J.R.; Arnold, J.G.; Srinivasan, R.; Ramanarayanan, T.S. APEX: A new tool for predicting the effects of climate and CO2 changes on erosion and water quality. In *Modelling Soil Erosion by Water*; Springer: Berlin/Heidelberg, Germany, 1998; pp. 441–449.

41. Wang, X.; Yen, H.; Liu, Q.; Liu, J. An auto-calibration tool for the Agricultural Policy Environmental eXtender (APEX) model. *Trans. ASABE* **2014**, *57*, 1087–1098.

42. Wang, X.; Gassman, P.; Williams, J.; Potter, S.; Kemanian, A. Modeling the impacts of soil management practices on runoff, sediment yield, maize productivity, and soil organic carbon using APEX. *Soil Tillage Res.* **2008**, *101*, 78–88. [CrossRef]

43. Cavero, J.; Barros, R.; Sellam, F.; Topcu, S.; Isidoro, D.; Hartani, T.; Lounis, A.; Ibrikci, H.; Cetin, M.; Williams, J. APEX simulation of best irrigation and N management strategies for off-site N pollution control in three Mediterranean irrigated watersheds. *Agric. Water Manag.* **2012**, *103*, 88–99. [CrossRef]

44. Zhang, B.; Feng, G.; Ahuja, L.R.; Kong, X.; Ouyang, Y.; Adeli, A.; Jenkins, J.N. Soybean crop-water production functions in a humid region across years and soils determined with APEX model. *Agric. Water Manag.* **2018**, *204*, 180–191. [CrossRef]

45. Jha, M.K.; Gassman, P.W.; Arnold, J.G. Water quality modeling for the Raccoon River watershed using SWAT. *Trans. ASABE* **2007**, *50*, 479–493. [CrossRef]

46. Moriasi, D.N.; King, K.W.; Bosch, D.D.; Bjorneberg, D.L.; Teet, S.; Guzman, J.A.; Williams, M.R. Framework to parameterize and validate APEX to support deployment of the nutrient tracking tool. *Agric. Water Manag.* **2016**, *177*, 146–164. [CrossRef]

47. Wang, X.; Williams, J.; Gassman, P.; Baffaut, C.; Izaurralde, R.; Jeong, J.; Kiniry, J. EPIC and APEX: Model use, calibration, and validation. *Trans. ASABE* **2012**, *55*, 1447–1462. [CrossRef]

48. Allen, R.; Pereira, L.; Raes, D.; Smith, M. Chapter 6-ETc-Single crop coefficient (KC). In *Crop. Evapotranspiration—Guidelines for Computing Crop Water Requirements—FAO Irrigation and Drainage Paper 56*; FAO—Food and Agriculture Organization of the United Nations: Rome, Italy, 1998.

49. Howell, T.A. Irrigation efficiency. In *Encyclopedia of Water Science*; Marcel Dekker: New York, NY, USA, 2003; pp. 467–472.

50. Altchenko, Y.; Villholth, K.G. Mapping irrigation potential from renewable groundwater in Africa—A quantitative hydrological approach. *Hydrol. Earth Syst. Sci. Discuss.* **2014**, *11*, 6065–6097. [CrossRef]

51. Mohan, S.; Simhadrirao, B.; Arumugam, N. Comparative study of effective rainfall estimation methods for lowland rice. *Water Resour. Manag.* **1996**, *10*, 35–44. [CrossRef]

52. Howsam, P.; Carter, R.C. Water Policy: Allocation and management in practice. In Proceedings of the International Conference on Water Policy, London, UK, 23–24 September 1996; p. 384.

53. FAO. *Guidelines for Land Use Planning*; Development Series 1; Soil Resources, Management and Conservation Service; Food and Agricultural Organization—FAO: Rome, Italy, 1993.

54. Chen, L.; Chan, C.M.; Lee, H.C.; Chung, Y.; Lai, F. Development of a decision support engine to assist patients with hospital selection. *J. Med. Syst.* **2014**, *38*, 59. [CrossRef] [PubMed]

55. Koczkodaj, W.W.; Mikhailov, L.; Redlarski, G.; Soltys, M.; Szybowski, J.; Tamazian, G.; Wajch, E.; Yuen, K.K.F. Important Facts and Observations about Pairwise Comparisons (the special issue edition). *Fundam. Inform.* **2016**, *144*, 291–307. [CrossRef]

56. Gebregziabher, G. Water Lifting Irrigation Technology Adoption in Ethiopia: Challenges and Opportunities. AgWater Case Study. Available online: http://awm-solutions.iwmi.org/Data/Sites/3/Documents/PDF/et-water-lifting-devices.pdf (accessed on 12 December 2012).

57. CSA. *[Ethiopia] Agricultural Sample Survey 2009/2010 (2002 E.C.) (September–December, 2009) Volume IV, Report on Area and Production of Crops Development*; Central Statistical Agency-Ministry of Finance and Economic Development: Addis Ababa, Ethiopia, 2010; Volume 446.

Article

Flooding Urban Landscapes: Analysis Using Combined Hydrodynamic and Hydrologic Modeling Approaches

Manoj K. Jha * and Sayma Afreen

Civil, Architectural, and Environmental Engineering, North Carolina A&T State University, Greensboro, NC 27411, USA; safreen@aggies.ncat.edu
* Correspondence: mkjha@ncat.edu

Received: 26 May 2020; Accepted: 11 July 2020; Published: 14 July 2020

Abstract: The frequency and severity of floods have been found to increase in recent decades, which have adverse effects on the environment, economics, and human lives. The catastrophe of such floods can be confronted with the advance prediction of floods and reliable analyses methods. This study developed a combined flood modeling system for the prediction of floods, and analysis of associated vulnerabilities on urban infrastructures. The application of the method was tested on the Blue River urban watershed in Missouri, USA, a watershed of historical significance for flood impacts and abundance of data availability for such analyses. The combined modeling system included two models: hydrodynamic model HEC-RAS (Hydrologic Engineering Center—River Analysis System) and hydrologic model SWAT (Soil and Water Assessment Tool). The SWAT model was developed for the watershed to predict time-series hydrograph data at desired locations, followed by the setup of HEC-RAS model for the analysis and prediction of flood extent. Both models were calibrated and validated independently using the observed data. The well-calibrated modeling setup was used to assess the extent of impacts of the hazard by identifying the flood risk zones and threatened critical infrastructures in flood zones through inundation mapping. Results demonstrate the usefulness of such combined modeling systems to predict the extent of flood inundation and thus support analyses of management strategies to deal with the risks associated with critical infrastructures in an urban setting. This approach will ultimately help with the integration of flood risk assessment information in the urban planning process.

Keywords: flood analysis; hydrologic modeling; hydrodynamic modeling; SWAT; HEC-RAS; flood zone delineation

---

1. Introduction

Over the years, the adverse effects of flooding have increased due to changing climate conditions and human interventions [1]. The major factors which lead communities to increased exposure of such flooding risks include urban expansion, changing demographic features within floodplains, changes in flood regime, and human intervention (social and economic developments) in the ecological system [2]. The hydro-meteorological catastrophes of such floods cannot be totally avoided, but the impacts and after-effects can be managed by developing the effective risk reduction and prevention strategies through applications of advanced geospatial tools and decision support systems [3]. Among the most effective ways of assessing the flood risk to people and infrastructures, one approach is the development and application of flood models which identify areas prone to flooding events and support risk analysis and mitigation processes [4]. Flood modeling has provided an indispensable tool to inform the development of the robust flood risk management strategies to avoid or mitigate the adverse impacts of floods on individuals, communities, and critical infrastructures such as transportation routes, hospitals, and others. A reliable flood model could alert the flood risk areas and warn the vulnerable population to relocate before the hazards take place. This will potentially alleviate the extent of devastation due to flooding and nullify causalities.

Flood modeling alludes to the process of transformation of rainfall into flood hydrographs which are then hydraulically translated into the depths of water at a spatial scale throughout the watershed [5]. Hydraulic models play an important role in determining flood inundation areas using sets of hydrodynamic equations. One of the major input data is the information on flood hydrographs at multiple locations as upstream and/or lateral boundary conditions. While these data can be obtained from observation data at gaging stations, these are often very limited or not available. Hydrologic rainfall-runoff models are thus frequently used, which when calibrated and validated to a reasonable accuracy, provide hydrograph information at desired locations. There are numerous studies that have independently evaluated the performance of hydrologic models [6–10] and hydrodynamic models [11–14] for their ability to perform the tasks they are developed for. New and improved algorithms have been continuously improving and evolving while capturing more robust simulation approaches and improved capabilities. Over the years, research efforts have been made to improve the numerical accuracy and computational efficiency of hydrodynamic flood models. However, the existing models are still computationally prohibitive for large-scale applications, especially in urban environments where high-resolution representation of complicated topographic features is necessary [14,15]. Similarly, hydrologic models can be computationally efficient in simulating hydrological processes but at the price of representing less detailed physical processes.

There have been several attempts combining hydrodynamic model with hydrological model which may compliment and overcome the shortcomings of either type of modeling approach. The integration of these models can be done various ways. External coupling uses the pre-acquired hydrographs from hydrological models as the upstream and/or lateral boundary conditions for the hydrodynamic models in flood routing analysis through complicated river network systems [16–18]. In the internal coupling method, governing equations of the hydraulic models and hydrological models are solved separately, with information at the shared boundaries updated and exchanged at each or several computational time steps [19]. Fully coupling of these models are not very well understood due to the complexities of reformulating and simultaneously solving governing equations in a single code base [14]. Other approaches include a hybrid method where a 2-D hydrodynamic model is combined with simplified unit hydrographs derived using variations of shallow-water equations [20–24] and integrated catchment models, suitable for flash flood modeling that simulate the complete hydrology and flow, generating runoff, leading to discharge, and then to flooding [25].

Combining hydrodynamic and hydrologic models for flood prediction and analysis is not new. However, the continuous modeling advances and the increase in computational resources over the years make it feasible to conduct flood simulations in high spatial resolution for flood risk assessment. In addition, scientific literature in combining of these two modeling approaches for urban flood simulation is limited [14], and thus underscores the need to continuously develop and apply robust models of improved capabilities for more efficient and accurate analysis. This study demonstrates the flood modeling and analysis method using advanced modeling tools of the present time via the combined or external coupling of hydrodynamic and hydrologic models. The hydrologic model, namely the Soil and Water Assessment Tool (SWAT) [26], was used to derive flow hydrographs at designated locations, which then fed into the hydrodynamic model, namely the Hydrologic Engineering Center's River Analysis System (HEC-RAS) [27] for flood prediction. Both models were independently calibrated and validated using sets of input databases, calibration techniques, observation data, and statistical performance evaluation methods. The combined application was used for flood simulations and the identification of the extent of inundation. The analysis provided the assessment of the impact of flood hazards by the identification of flood risk zones and the threatened infrastructures. The approach was applied in the Blue River Watershed in Missouri, USA, which has historic significance with respect to frequent severe floods. The watershed provides a rich database of observation data developed over the years. The combined modeling system provides crucial flood risk information necessary for the development of an accurate and reliable forecasting system for assist in evacuation, relief operation route, cost estimation of the damaged properties, and other pertinent information.

2. Materials and Methods

The method included development of two mathematical models: SWAT and HEC-RAS. The SWAT model was developed via Geographic Information System (ArcGIS) interface of the model, called ArcSWAT, using a set of spatial datasets including topography data (digital elevation model), land use data (National Land Cover Database), and soil types and soil characteristics data (State Soil Geography Database), as well as time-series daily dataset on meteorological parameters, including precipitation, maximum and minimum temperature, wind speed, solar radiation, and relative humidity. The model was calibrated for the overall watershed hydrological water balance followed by monthly streamflow at a gaging location at the watershed outlet by comparing model simulated values with the observation data collected at the gaging site. Once the model is calibrated and validated with satisfactory statistical performance measures, it was then used to develop a series of simulated streamflow hydrographs to be used as an input to the HEC-RAS model.

The HEC-RAS model was developed for river segments within the watershed. The geometric data and the Manning's roughness coefficient values (n) were established for the modeling setup using ArcGIS interface of the model, called HEC_geoRAS. It was then calibrated and validated using the past flood data collected from the USGS gaging stations within the study area.

The flow chart in the Figure 1 portrays different step of the processes performed in this study. Based on the streamflow input from calibrated SWAT model, the calibrated HEC-RAS model predicts flood levels and the extent of the flood in the surrounding landscapes. Further analysis was conducted to identify vulnerabilities of critical infrastructures including hospitals, railroads, airports, and transportation routes by examining the proximities of these infrastructures from the flood zones.

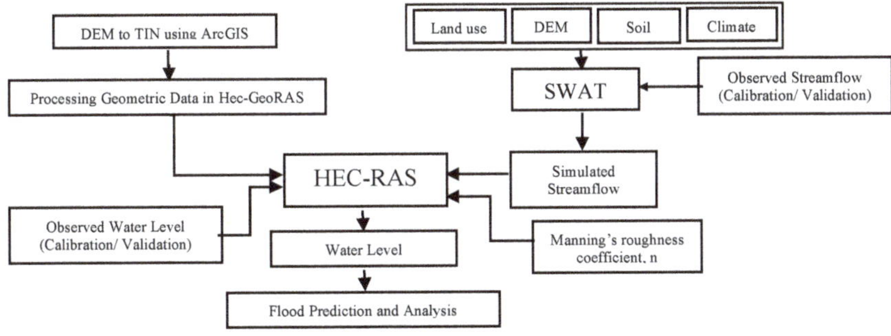

Figure 1. Schematic of data and models for flood prediction and analysis.

2.1. Study Area

The Blue River also known as Big Blue River is a part of tributaries of Missouri River located in Kansas City, Missouri (Figure 2). The Blue River watershed extends from the south of Johnson County in Kansas State into the State of Missouri and drains an area of 658.9 km^2 into the Missouri River in Kansas City, Jackson County, Missouri. The Blue River Watershed spreads over roughly one-half of the Kansas City metropolitan area south of the Missouri River. The watershed course through two states (Missouri and Kansas), four counties (Johnson and Wyandotte in Kansas; Jackson and Cass in Missouri), and 11 municipalities [28]. The Blue River is 39.8 mile (64.1 km) long stream, and the mouth of the river is at 221 feet elevation in the east of Johnson County near the borders of the states of Kansas and Missouri. The percentage of Blue river watershed within the state of Missouri is about 46%, which is within the Kansas City metropolitan area. The area is moderately to highly developed and contain a mix of residential and commercial structures and is subjected to flooding every year due to urban development, dense soils, and the configuration of the Blue River basin [29]. The lower part of the watershed is primarily industrial, whereas the middle and upper part are rapidly being converted to residential areas [30]. Due to the flood sensitive nature of this river zone, U.S. Geological Survey (USGS) has been studying this area closely and acquired an extensive dataset over the time. The abundance of data in this location was very helpful in accurately calibrating the mathematical models for flood prediction and the analysis objective of this study.

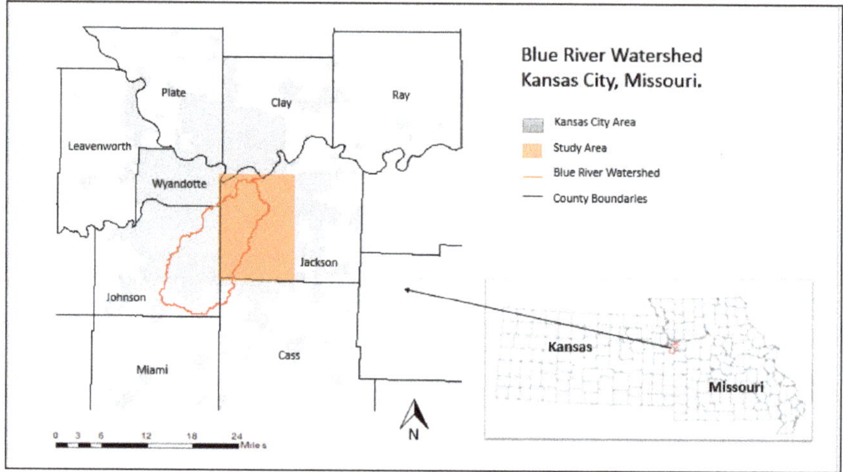

Figure 2. Blue River Watershed in Kansas City, Missouri, USA.

2.2. Data Collection

The development of the combined modeling system required extensive data collection from various sources. Table 1 lists some of the major data types and their sources. Subsections below provided more specific information on data collection efforts.

Table 1. Datasets and their sources used for creating the combined modeling system.

Dataset	Source
Digital Elevation Model (DEM)	United States Geological Survey (USGS)–The National Map
Streamflow	United States Geological Survey (USGS)
Gage Height	United States Geological Survey (USGS)
Land Use	National Land Cover Database (NLCD)
Climate Data	National Oceanic & Atmospheric Administration (NOAA)
Soil Classification	State Soil Geography Database (STATGO)

2.3. Hydrologic Model Overview–SWAT

SWAT is a river basin scale model developed to predict the impact of land management practices on water, sediment and agricultural chemical yields in large complex watershed with varying soils, land use and management conditions over extended periods of time [6,20]. SWAT is a long-term yield model extensively used to simulate watersheds on multiple spatial–temporal scales including hydrological processes [7,9,31,32], fate and transport of sediment and nutrients [33–35], land use change [36], climate change [37–43], and others.

The major inputs required to develop a SWAT model are topographical data which are used to define stream network and delineate a number of subwatersheds; land use data, soil data, and slope information to delineate each subwatershed into hydrologic response units (HRUs) which represents unique combination of land use, soil types, and slope; (3) the daily time-series information on meteorological parameters, and (4) the model's inbuilt databases and initialization assumptions. The outputs include spatiotemporal time-series data on water balance components, streamflow, sediment and nutrient loadings, and others.

2.3.1. Modeling Setup and Watershed Delineation

The Blue River Watershed was delineated using stream generation functionality in ArcGIS based on the supplied 10 m resolution DEM projected in Northern America Datum NAD_1983 UTM zone 15N. The delineated subwatersheds (Figure 3) were further subdivided into multiple lumped units within each subwatershed. These lumped units are called HRUs, a unique combination of land use, slope, and soil types. An HRU represents a percentage of a sub-watershed area and not spatially identified within a subwatershed. All water balance calculations and modeling simulations are conducted at the HRU level. Outputs from each HRU within a subwatershed are aggregated at the subwatershed level which are then routed through the streams leading to the next downstream subwatershed. Outputs from each subwatershed are subsequently routed all the way to the watershed outlet on a daily basis. Muskinghum method was used in the hydrologic routing process. Other methods include Curve-Number approach for flow generation, and Penman-Moneith method for the estimation of evapotranpiration.

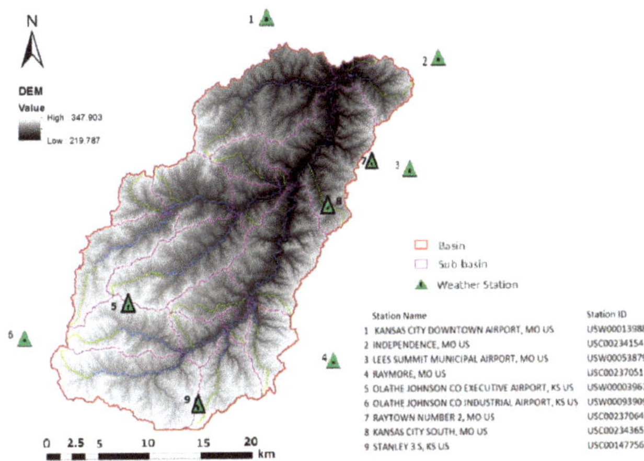

Figure 3. Delineation of Blue River Watershed and location of NOAA weather stations.

The land use data was obtained from National Land Cover Dataset (NLCD) from 2011 (https://www.mrlc.gov/nlcd11_data.php). Classification of the land use data was found to cover dense urban areas (48%), developed open area (20%), pasture/hay (13.5%), forest (9%), cultivated crops (7%), grassland (1%), open water (0.6%), wetland (0.5%), shrub (0.2%), and barren land (0.2%). The soil data source was State Soil Geographic (STATSGO) database (https://catalog.data.gov/dataset/statsgo) which was already included in the ArcSWAT inbuilt datasets. Figure 4 presents reclassified land use, soil types and slope categories used in HRU delineation. The time-series meteorological information was obtained for 9 weather stations located in and around the watershed (Figure 3) using data download function at the NOAA-NCDC website (https://www.ncdc.noaa.gov/cdo-web/datatools/findstation).

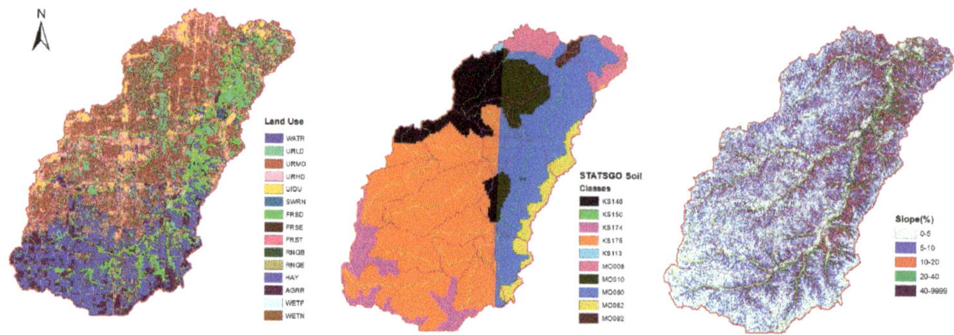

Figure 4. Reclassified land use, soil and slop data used in the HRU delineation.

2.3.2. SWAT Simulation and Calibration/Validation Approach

The SWAT modeling setup was executed on a daily time step for 8 years of simulation from 2010–2017 (2-year warm-up, 4-year calibration, and 2-year validation period). Calibration and validation of the SWAT model was performed using SWAT's Calibration and Uncertainty Program, SWAT-CUP [44]. This autocalibration tool can perform sensitivity analysis, calibration, validation, and uncertainty analysis. Sensitivity analysis of the model's hydrologic parameters were conducted and ten parameters were identified as the most sensitive. There are runoff curve number (CN2), soil evaporation compensation factor (ESCO), water holding capacity of the soil (SOL_AWC), plant uptake compensation factor (EPCO), groundwater revap coefficient (GW_REVAP), base flow alpha factor (ALPHA_BF), threshold depth of water in the shallow aquifer required for return flow (REVAPMN), groundwater delay (GW_DELAY), surface runoff lag coefficient (SURLAG), and threshold depth of water in the shallow aquifer required for return flow to occur (GWQMN). The details of these model parameters can be found in the User's Manual. In the calibration process, defaulted values of these parameters were adjusted within their permissible ranges to a final calibrated value after comparing simulated results with the observations with acceptable performance measures tested through statistical procedures [39]. The autocalibration tool identified the best fitted values of all ten parameters while fitting the monthly comparison of simulated flow values with the observations from gaging stations at the watershed outlet. Statistical evaluation was conducted using four indicators: coefficient of determination (R^2), Nash–Sutcliff's efficiency (NSE), percentage bias (PBIAS), and RMSE standard deviation ratio (RSR). Calibration process concluded with satisfactory performance in visual comparison and acceptable statistical comparisons. During the validation process, the model was executed with already defined value of calibration parameters (no further), followed by the same statistical evaluation as that of the calibration duration.

2.4. Hydrodynamic Model Overview–HEC-RAS

HEC-RAS [21] can perform one and two-dimensional hydrodynamic calculations for a full network of natural and constructed channels. The major capabilities of HEC-RAS are user interface, hydraulic analysis components, data storage and management, graphics and reporting, and RAS Mapper. The HEC-RAS system accommodates several river analysis components for steady and unsteady flow water surface profile computations, movable boundary sediment transport computations and water quality analysis. Hydrodynamic equations calculate water surface elevations at all locations of interest for a given peak flood. The major data inputs are river geometric cross-section data, river floodplain data (length, elevation), the distance between successive river cross-sections, manning roughness coefficient values (n) for the land use type covering the river and the floodplain area, and boundary conditions (flow hydrograph and normal depth). Under steady flow, the boundary conditions are a discharge upstream and a stage downstream. The model proceeds to calculate stages throughout the interior points, keeping the discharge constant in space. Under unsteady flow, a discharge hydrograph is the upstream boundary and a discharge-stage rating at the downstream boundary. The model calculates discharges and stages throughout the interior points. Unsteady flow simulation uses the Saint-Venant equations or the diffusion wave equations using an implicit finite volume algorithm. The outputs from the HEC-RAS model include water surface elevations, rating curves, hydraulic properties (energy grade line slope, elevation, flow area, velocity), and visualization of the extent of flooding.

The steady flow simulation based on a peak flow discharge throughout the river line represents the water flow without any change over time. It consists of flow regime, discharge information and boundary condition. Multiple profiles can be created with different discharge values. The unsteady flow simulation is developed with a series of discharge data with respect to the time of occurrence. The data required for the unsteady flow simulations include boundary conditions (external and internal) and initial conditions.

The calibration of the model was initiated by calibrating for the steady flow simulation followed by the calibration of the unsteady flow simulations. The model was calibrated for a peak flow event at five USGS gaging stations (Figure 5) with adjustments in the parameters such as Manning's n value and required boundary conditions [21]. The upstream boundary condition was a flow hydrograph and the downstream boundary condition was normal depth for steady state simulation. The HEC-RAS was executed to develop water level data which were compared with observed water elevations. After the calibration, the model was validated for two other flood events at all five stations based on the calibrated parameters.

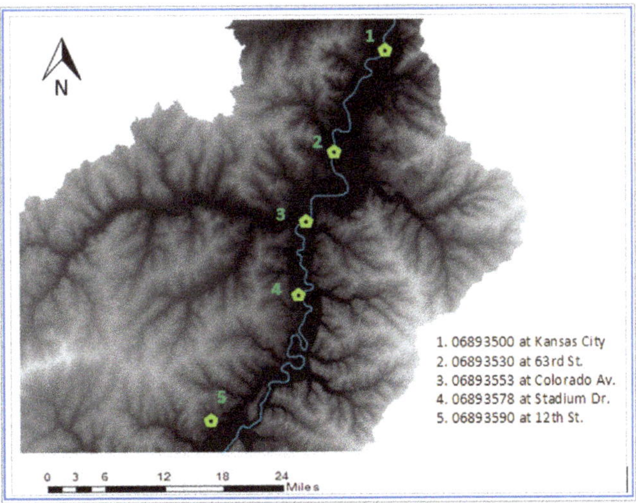

Figure 5. Location of USGS Gaging Stations used in the calibration.

3. Results and Discussion

3.1. Clibration and Validation of the Hydrolgic Model

The SWAT model developed for the Blue River Watershed was calibrated using the automated calibration technique (SUFI-2) for flow by comparing simulated values with the observations at the watershed outlet (USGS 06893500, Blue river at Kansas City, MO, USA). Table 2 lists all parameters used in the calibration process with their permissible ranges and the final fitted values after the calibration.

Table 2. List of parameters used for calibration with their ranges and the fitted value.

Parameter	Description	Range	Fitted Value
CN2	Curve Number	−15%–15%	−5.34%
EPCO	Plant Uptake Compensation Factor	0.01–1	0.73
SOL_AWC	Water Holding Capacity of Soil	−0.04–0.04	−0.025
GW_REVAP	Groundwater Revap Coefficient	0.02–0.2	0.05
ALPHA_BF	Base Flow Alpha Factor	0.05–0.8	0.1
REVAPMN	Threshold Depth, Percolation to deep aq.	0–500	455
ESCO	Soil Evaporation Compensation Factor	0.75–0.95	0.81
GW_DELAY	Groundwater Delay	0–500	476
GWQMN	Threshold Depth, Return flow to occur	0–1000	868
SURLAG	Surface Runoff Lag Coefficient	1–8	6.7

Figure 6 shows the comparison of simulated versus observed streamflow at the watershed outlet using monthly data. The comparison seems to match well except for slight underprediction of peaks. The hydrograph seems to follow very close for its recession, baseflow and other patterns. Table 3 provides values of statistical measures for both calibration and validation periods. Overall, these values show a strong correlation of the simulated streamflow with the observation. Thus, it can be concluded that the SWAT model was well-calibrated to simulate streamflow with reliable performance in the Blue River Watershed. The calibrated model output was used to generate discharge (streamflow) data at several locations within the watershed to be used as input for the HEC-RAS model.

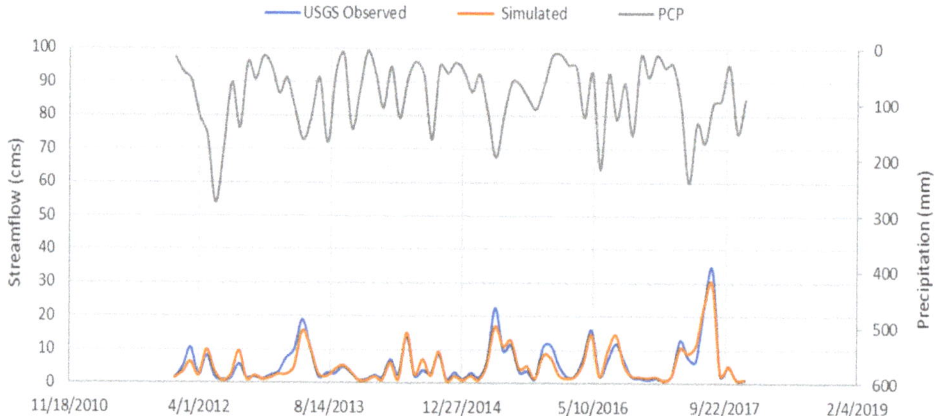

Figure 6. Monthly comparison of simulated and observed streamflow data for the calibration (2012–2015) and validation (2016–2017) periods along with precipitation data for the entire range (cms: m^3/s).

Table 3. Statistical Evaluation of the calibration & validation of the Blue River Watershed.

Statistical Test	Calibration Period 2012–2015	Validation Period 2016–2017	Acceptable Range [45]
NSE	0.83	0.92	NSE > 0.50
PBIAS (%)	9.40	3.00	PBIAS < ±25%
RSR	0.41	0.28	RSR < 0.70
R^2	0.84	0.93	R2 > 0.5

3.2. Clibration and Validation of the Hydrodynamic Model

The hydrodynamic model developed for the Blue River by HEC-RAS was calibrated and validated at the five USGS gaging stations located on the River (Figure 5). Model simulated water surface elevations were compared with the observed water surface elevations at the USGS gages. The Manning's roughness coefficient (n) values were adjusted until the simulated values match closely with the values at USGS gages. The calibration was performed for the flood event of 17 May 2015 and the results are presented in Table 4. The results show that the difference between the observed and simulated values were very minimal and thus justify the model's ability to simulate water surface levels. The statistical evaluation using two performance measures NSE and R^2 yielded a strong correlation with value of 0.989 and 0.98 respectively. The validation process was conducted for two peak events: the floods on 27 April 2016 and 22 September 2017. The difference in observed and simulated water surface elevations were small and therefore the results are considered satisfactory which is portrayed in Table 5. It can be concluded that the HEC-RAS model developed for the Blue River performed very well to simulate water surface elevations.

Table 4. HEC-RAS Model Calibration for the flood event of 17 May 2015.

USGS Station	Flow (m³/s)	Simulated Stage (m)	Observed Stage (m)	Difference (m)
06893500	298	7.8	7.7	−0.06
06893530	268	7.0	7.0	0.05
06893553	251	6.2	6.2	−0.01
06893578	229	6.0	6.0	−0.14
06893590	178	5.8	5.8	0.03

Table 5. HEC-RAS Model Validation for flood events of 27 April 2016 and 22 September 2017.

USGS Station	Event 4/27/2016				Event 9/22/2017			
	Flow (m³/s)	Simulated Stage (m)	Observed Stage (m)	Difference (m)	Flow (m³/s)	Simulated Stage (m)	Observed Stage (m)	Difference (m)
06893500	239	7.3	7.0	−0.34	1203	15.4	14.1	−1.31
06893530	1171	16.1	14.9	−1.19	237	6.6	7.2	0.59
06893553	1162	13.8	14.4	0.63	236	6.5	7.1	0.61
06893578	1154	12.6	10.7	−1.88	235	6.5	6.3	−0.23
06893590	1138	11.2	9.5	−1.63	233	6.3	5.5	−0.82

3.3. Flood Inundation Mapping

Accurate prediction of the flood inundation area for a given flood event is necessary for risk mitigation strategies. Over the last few decades, there have been vast improvements in flood inundation modeling [46]. While empirical methods are considered adequate for flood monitoring and post-disaster assessment, hydrodynamic models are critical to represent detailed flow dynamics to investigate impacts of management strategies such as dam break, flash floods, etc. Simplified conceptual models are usually adopted for probabilistic flood risk assessment on a large floodplain with well-defined channels. Different modeling approaches produce different predictions highlighting the uncertainty associated with the modeling practices, which is mainly generated by uncertainty in the design flow, terrain elevations, water surface elevations, and accuracy of the techniques used for mapping the inundation area [47].

In this study, the flood inundation area was developed using ArcGIS based on the HEC-RAS simulation of desired flood event. The pseudo-validation of the developed inundation map was conducted by comparing it with inundation maps already developed by the USGS which was available to view/download from the USGS Flood Inundation Mapper (https://wimcloud.usgs.gov/apps/FIM/FloodInundationMapper.html). Figures 7 and 8 show the comparison between inundation maps created by HEC-RAS simulation (right figures) with the USGS inundation maps (left figures) at two separate locations. The comparison was done visually by comparing important features along the Blue River. The maps fairly show comparable zones of inundated area in both cases. It is important to note that the discharge data used by HEC-RAS in developing the inundation extent was based on the "simulated" discharge data from the hydrologic model, which may have contributed greatly to the disagreements between the two maps. Moreover, the comparison was against the another simulated map as explained in the disclaimer by the USGS (https://fim.wim.usgs.gov/fim/) which states that "the flood boundaries shown were estimated based on water stages (water-surface elevations) and streamflows at selected USGS streamgages. Water-surface elevations along the stream reaches were estimated by steady-state hydraulic modeling, assuming unobstructed flow, and using streamflows and hydrologic conditions anticipated at the USGS streamgage(s)".

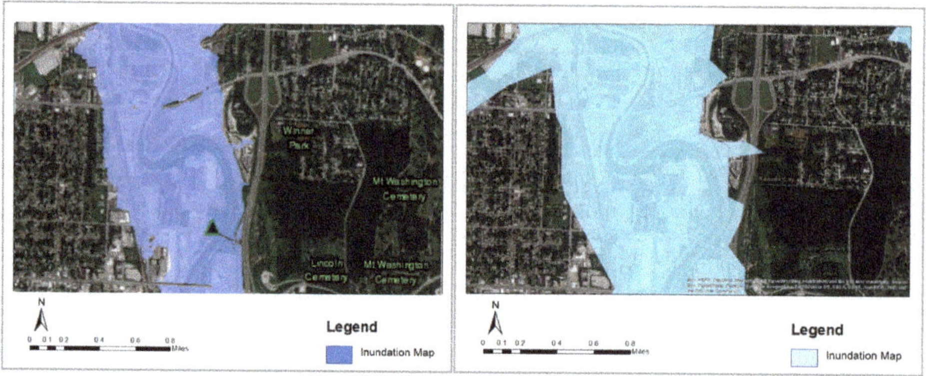

Figure 7. Comparison of the Inundation Map created by the HEC-RAS simulation (**right-side**) with the USGS (**left-side**) generated Inundation Map beside the Winner Park in Kansas City, MO.

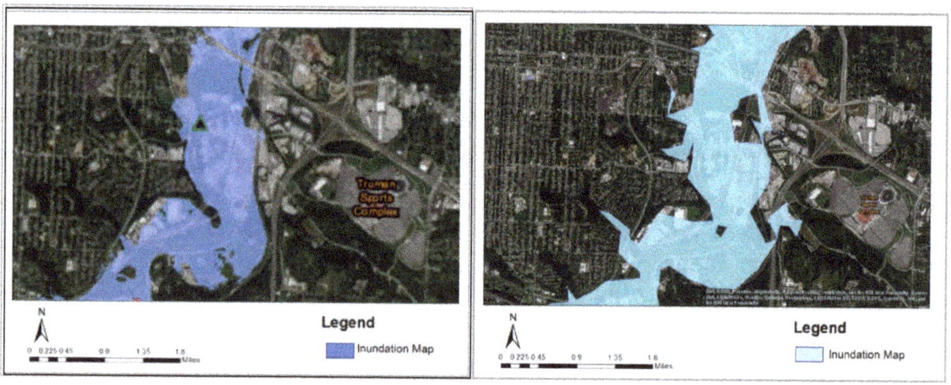

Figure 8. Comparison of the Inundation Map created by the HEC-RAS simulation (**right-side**) with the USGS (**left-side**) generated Inundation Map beside the Truman Sports Complex in Kansas City, MO.

3.4. Vulnerability Assessment on Infrastructures

Vulnerability assessment is an essential part of the flood management and preparedness process to reduce the impact. It requires an in-depth analysis of many factors including location of critical infrastructures such as hospitals, transportation routes and density of the population in order to increase the effectiveness of emergency plans. Indicators of flood hazard generally include the flood extent, water depth, flow velocity, duration, propagation of waterfront, and the rate at which the water rises [48]. These parameters are then linked with the economic damages and other vulnerability assessment. There are many studies linking inundation extent to determine economic losses or risks for planning purposes such as insurance, etc. [49]. The vulnerability criterion focused on human stability (not economic values) has also been analyzed using slipping, toppling, and drowning as indicators of human stability [50]. A flood modeling simulation in an urban area used inundation maps to analyze transport accessibility and human safety on pedestrians and drivers for its implications on emergency routes and service areas [51,52]. A comparative study of hydraulic models evaluated their capabilities for estimations of the vulnerability assessment to capture the uncertainties in the prediction [53].

In this study, we present the vulnerability assessment in terms of critical infrastructures being exposed to floodwaters by proximity to flood inundation extent over the study area. The analysis was based on flood event of May 2015. The infrastructures selected for vulnerability assessment purposes were local hospitals, transportation routes, airport facilities, and railroad networks. These infrastructures are very crucial for emergency responses such as for mitigation, preparedness, recovery, and response. For example, emergency response teams could use the inundation maps to optimize their routes to the flood affected locations, avoiding the inundated transportation routes. The inundation maps could also assist in the allocation of recovery resources from the high-risk zones following a flood event. Inundation maps could be created assuming a future storm event causing a flood, and therefore highly threatened flood zones could be alarmed ahead of time, thereby saving lives and resources.

3.4.1. Impact of Inundation on Local Hospitals

Hospitals are one of the major locations highly prioritized in the disaster mitigation process. Figure 9 depicts four hospitals that could be threatened due to similar flood situation like as May 2015. One of the four risked hospitals was identified to be almost under inundation and the rest could be impacted with an increase of a few units of water level caused by a more hazardous flood. The surrounding hospitals could be indirectly affected due to the closure of the nearby transportation routes. This vulnerability identification could help the management authorities warn the hospitals listed under the adverse impact, ahead of any upcoming hazards. The vulnerable hospitals showed in Figure 9 are listed in Table 6 with their distance from the inundation area. To understand the different levels of flood vulnerability, a ranking is given to the hospitals with respect to the distance of the hospitals from the flood extent at their respective locations.

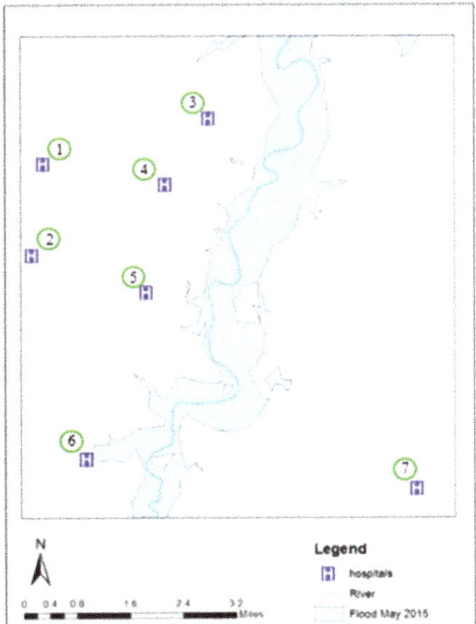

Figure 9. Location of hospitals in and around the inundated area due to Flood of May 2015.

Table 6. List of hospitals vulnerable to the flood of May 2017 in Kansas City, Missouri.

Hospital Location (Figure 10) and Names	Distance from Inundation Area (km)	Vulnerability Rank
1H. Seton Center Safety Net Clinics	3.3	7
2H. Samuel U. Rodgers South Therapeutic Intervention Center-Substance Abuse	2.9	5
3H. Samuel U. Rodgers McCoy Elementary School Dental Clinic	0.8	3
4H. Kansas City Free Health Clinic-Eastside	1.0	4
5H. Veterans Affairs Medical Center	0.7	2
6H. Swope Health Services-Central	0.1	1
7H. Two Rivers Psychiatric Hospital	3.0	6

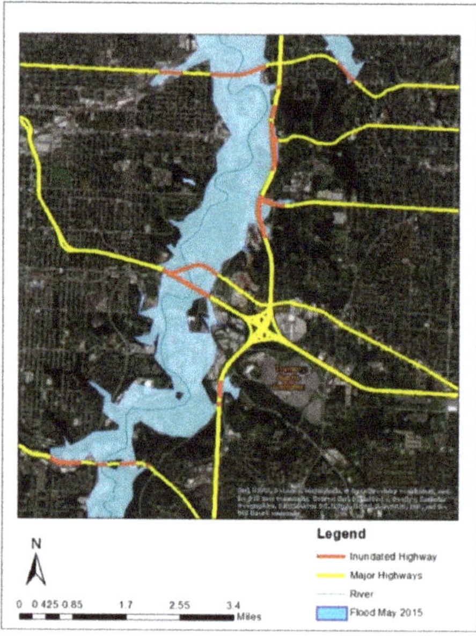

Figure 10. Location of major transportation routes under inundation due to the flood of May 2015.

3.4.2. Impact of Inundation on Transportation Routes

Intense precipitation is the foremost cause of weather-related disruption to the transportation sector [39]. It can cause severe damage to an area by obstructing the movement of people and goods, hampering social and economic functionality. The flooding on major transportation routes, like interstates and state highways, cut off the flooded zone's communication with the surrounding area which also delays the emergency management processes. Figure 10 shows the transportation routes that are directly affected due to the flooding scenario modeled for May 2015 flood. Parts of the interstate I70, Blue Parkway, Highway I435, Highway US 40, and Independence Avenue are found to be under the impact of inundation caused by the flood of May 2015 as simulated by HEC-RAS.

3.4.3. Impact of Inundation on Airport

Figure 11 shows the threatened location of Airports due to flood of May 2015. One of the airport facilities will be directly affected by the flood and the other one is very close by the inundated regions. A ranking is given to airports for flood vulnerability with respect to the distance of the facility from the inundation map at their respective locations (Table 7).

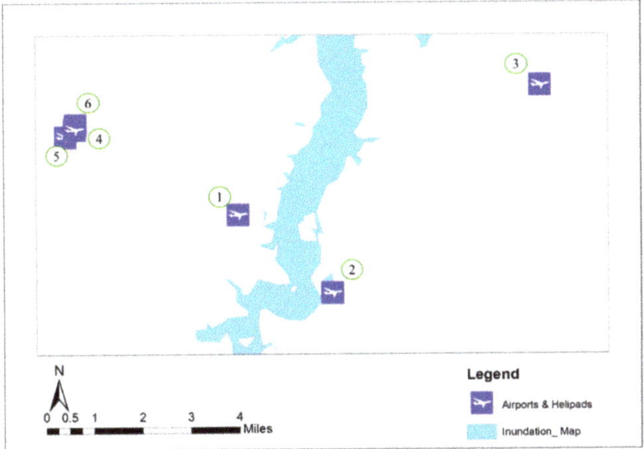

Figure 11. Location of airport under inundation due to the Flood of May 2015.

Table 7. List of airports vulnerable to the flood of May 2017 in Kansas City, Missouri.

Airport Location (Figure 12) and Names	Distance from Inundation Map (m)	Vulnerability Rank
1. VA Medical Center Heliport	670	2
2. Police Department Helipad Main Facility	126	1
3. Independence RGNL Health Center Heliport	3041	3
4. Truman Medical Center West Heliport	5316	4
5. Children's Mercy Hospital Heliport	5465	6
6. Bert Walter Berkowitz Heliport	5341	5

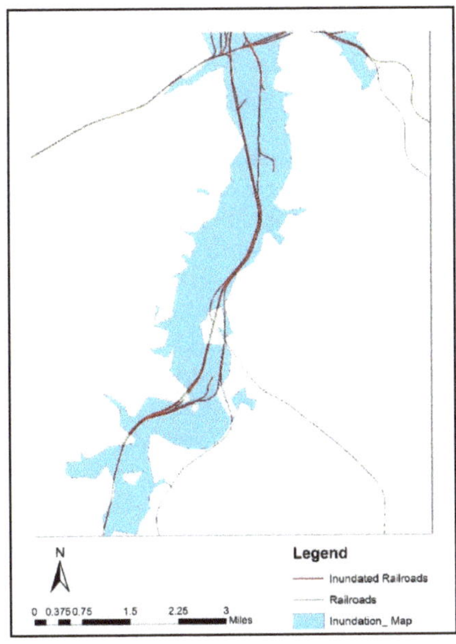

Figure 12. Location of railroad routes under inundation due to the Flood of May 2015.

3.4.4. Impact of Inundation on Railroad Facilities

The railroad is one of the most used routes in big cities and metropolitan areas. Inundated railroads could cause fatal accidents that would affect huge numbers of people travelling in the trains. Figure 12 shows the location of railroad routes under inundation due to the flood of May 2015 as simulated by HEC-RAS. Railroads that are subjected to inundation will result in obstruction of the whole rail route within and surrounding the city. Table 8 provides the coverage of railroads under the flood.

Table 8. List of railroads vulnerable to the flood of May 2017 in Kansas City, Missouri.

Name of Railroad	Length Under Inundation (m)
BNSF RR	3014
KCS RR	3844
KCT RR	2390
Missouri Central RR	362
Private RR	333
UP RR	5750

4. Discussion and Conclusion

This study presents a systematic approach of combining hydrodynamic model HEC-RAS with hydrologic model SWAT in delineating flood inundation zones and subsequently assessing the vulnerability of critical infrastructures in the Blue River Watershed in Kansas City, Missouri. Both models were independently calibrated and validated using various datasets and proven strategies. The HEC-RAS flood simulation model was found to be suitable in simulating flood events and spatially depicting the vulnerability of the region towards a hazard event in terms of inundation extent, whereas SWAT was proven to be a powerful tool in generating simulated flood hydrographs at desired locations. The models developed can be said to have generated reliable quantified output based on the statistical evaluation results. This study approach provides quantified information on the hydrologic modeling, hydrodynamic modeling, and flood prediction and analysis for flood management strategies.

The catastrophic possessions of flood disaster could be mitigated by integrating scientifically reliable information with the flood inundation map developed using this study approach. Vulnerability assessment approach used in this study for identifying and providing a vulnerability rank based on proximity to flood area is a simple yet powerful approach. It not only identified most to least vulnerable critical infrastructures, but also provided enough information for flood preparedness processes that could significantly reduce the impact. The approach could easily be extended for the vulnerability evaluation of other infrastructures in order to estimate economic losses, navigation route of people including high density area, and other region-specific important factors. Moreover, futuristic higher magnitude flood events can be simulated to assess magnified vulnerability and associated risks. Land use planning decisions could be made based on the flood inundation map which indicates the floodplains. Following such approaches will help save lives and resources at the same time, and provide a proven and more accurate way to contest the uncertainties of the natural events causing flood.

The flood modeling system presented in this study is an integrated system to stakeholders to investigate potential mitigation options and strategies in response to expected flooding scenarios. The use of hydrologic model in flood modeling proves very useful in studying alternative "what if" scenarios such as impacts of projected land use changes, climate variabilities, urban planning strategies and others. For all plausible scenarios, a well-calibrated hydrologic model of the region can easily simulate new conditions and yield changes in flow hydrographs at desired locations, which can then be translated into flood depths over the region using hydrodynamic models. A previous flood modeling attempt in Kansas River basin, close to the study watershed, used hydrologic model HEC-HMS (Hydrologic Modeling System) to generate estimates of peak flows for design storm for different land use scenarios [4]. The output was then used to execute the HEC-RAS model for estimates of water elevations and flood inundation extents for those design storms and land use scenarios. The results provided useful information, however the study was designed at a macro scale of change which does not necessarily reflect the flooding impacts at smaller scale.

Added benefits of this combined modeling presented in this study system also includes the flexibility of hydrodynamic modeling for testing flood reduction or mitigation strategies through channel modifications and other best management practices within the floodplains areas. Such a modeling system also enables the assessment and determination of vulnerable areas that will not be able to receive effective adaptation solutions, which then calls for drastic measures to mitigate flood-prone impacts.

Such modeling application also comes with several limitations, including the availability, resolution, and accuracy of the data for the development, calibration, and validation of the models, the integration methods such as external coupling approach used in this study, flexibility provided by the statistical performance measures for the approval of a robust model, and ability to replicate/simulate best management practices with a degree of accuracy to support flood mitigation and adaptation options. As such, in the application presented in this study, major limitations in using hydrologic model may include (a) limited accuracy in model calibration and validation: resolution of the input data and limited set of observation data, e.g., calibrating only for the monthly flow and only at the watershed outlet, and (b) simulated data to be exported as input to another model: calibrated models produced simulated hydrographs to be used as input boundary conditions in hydrodynamic modeling. Similarly, sources of uncertainties in using hydrodynamic modeling may include input data quality of topography and surface roughness characterization as it affect both flow area and velocity [54,55]. The role of topography on flood studies has been discussed in many past studies [56], but the role of surface roughness has received less attention. A recent study exhibits the sensitiveness of surface roughness and highlights the source of uncertainties in flood modeling studies [57]. Reducing the uncertainty in surface roughness will greatly enhance the calculation of flood extent on landscapes. It is also noteworthy to mention that, while surface roughness plays an important role in simulating accurate flow hydrodynamics in both the channel and floodplain, Manning's n is not viewed as important (less-sensitive parameter) in simplified hydrological modeling.

Moreover, the errors in the simulation results from the combined modeling system of flood analysis arise from various sources of uncertainties, as discussed above, which probably propagates in an unknown and non-linear fashion. The next level of analysis should shed light on the assessment and quantification of these errors and how these propagate through the modeling system.

Author Contributions: Conceptualization, M.K.J.; methodology, M.K.J. and S.A.; software, M.K.J.; validation, M.K.J. and S.A.; formal analysis, S.A.; investigation, S.A. and M.K.J.; resources, S.A.; data curation, S.A.; writing—original draft preparation, S.A. and M.K.J.; writing—review and editing, M.K.J. All authors have read and agreed to the published version of the manuscript.

Funding: This research received no external funding.

Conflicts of Interest: The authors declare no conflict of interest.

References

1. Bronstert, A. Floods and climate change: Interactions and impacts. *Risk Anal.* **2003**, *23*, 545–557. [CrossRef]
2. Dang, N.M.; Babel, M.S.; Luong, H.T. Evaluation of food risk parameters in the Day River Flood Diversion Area, Red River Delta, Vietnam. *Nat. Hazards* **2011**, *56*, 169–194. [CrossRef]
3. Khan, S.I.; Hong, Y.; Gourley, J.J.; Khattak, M.U.; De Groeve, T. Multi-Sensor Imaging and Space-Ground Cross-Validation for 2010 Flood along Indus River, Pakistan. *Remote Sens.* **2014**, *6*, 2393–2407. [CrossRef]
4. Yuan, Y.; Qaiser, K. *Floodplain Modeling in the Kansas River Basin Using Hydrologic Engineering Center (HEC) Models: Impacts of Urbanization and Wetlands for Mitigation*; US Environmental Protection Agency: Washington, DC, USA, 2011; pp. 1–32.
5. Ramírez, J.A. Prediction and modeling of flood hydrology and hydraulics. In *Inland Flood Hazards: Human, Riparian and Aquatic Communities*; Ellen, W., Ed.; Cambridge University Press: London, UK, 2010; Chapter 11.
6. Arnold, J.G.; Moriasi, D.N.; Gassman, P.W.; Abbaspour, K.C.; White, M.J.; Srinivasan, R.; Santhi, C.; Harmel, R.D.; Van Griensven, A.; Van Liew, M.W.; et al. SWAT: Model Use, Calibration, and Validation. *Trans. ASABE* **2012**, *55*, 1491–1508. [CrossRef]

7. Jha, M.K. Evaluating Hydrologic Response of an Agricultural Watershed for Watershed Analysis. *Water* **2011**, *3*, 604–617. [CrossRef]
8. Panagopoulos, Y.; Gassman, P.W.; Jha, M.K.; Kling, C.L.; Campbell, T.; Srinivasan, R.; White, M.; Arnold, J.G. A refined regional modeling approach for the Corn Belt—Experiences and recommendations for large-scale integrated modeling. *J. Hydrol.* **2015**, *524*, 348–366. [CrossRef]
9. Baffaut, C.; Dabney, S.M.; Smolen, M.D.; Youssef, M.A.; Bonta, J.V.; Chu, M.L.; Guzman, J.A.; Shedekar, V.S.; Jha, M.K.; Arnold, J.G. Hydrologic and Water Quality Modeling: Spatial and Temporal Considerations. *Trans. ASABE* **2015**, *58*, 1661–1680.
10. Guzman, J.A.; Shirmohammadi, A.; Sadeghi, A.M.; Wang, X.; Chu, M.L.; Jha, M.K.; Parajuli, P.B.; Harmel, R.D.; Khare, Y.P.; Hernandez, J.E. Uncertainty Considerations in Calibration and Validation of Hydrologic and Water Quality Models. *Trans. ASABE* **2015**, *58*, 1745–1762.
11. Shustikova, I.; Domeneghetti, A.; Neal, J.C.; Bates, P.; Castellarin, A. Comparing 2D capabilities of HEC-RAS and LISFLOOD-FP on complex topography. *Hydrol. Sci. J.* **2019**, *64*, 1769–1782. [CrossRef]
12. Patel, D.; Ramirez, J.A.; Srivastava, P.K.; Bray, M.; Han, D. Assessment of flood inundation mapping of Surat city by coupled 1D/2D hydrodynamic modeling: A case application of the new HEC-RAS 5. *Nat. Hazards* **2017**, *89*, 93–130. [CrossRef]
13. Pinos, J.; Timbe, L. Performance assessment of two-dimensional hydraulic models for generation of flood inundation maps in mountain river basins. *Water Sci. Eng.* **2019**, *12*, 11–18. [CrossRef]
14. Liu, Z.; Zhang, H.; Liang, Q. A coupled hydrological and hydrodynamic model for flood simulation. *Hydrol. Res.* **2019**, *50*, 589–606. [CrossRef]
15. Hunter, N.M.; Bates, P.D.; Horritt, M.S.; Wilson, M. Simple spatially-distributed models for predicting flood inundation: A review. *Geomorphology* **2007**, *90*, 208–225. [CrossRef]
16. De Paiva, R.C.; Collischonn, W.; Tucci, C.E. Large scale hydrologic and hydrodynamic modeling using limited data and a GIS based approach. *J. Hydrol.* **2011**, *406*, 170–181. [CrossRef]
17. Kim, J.; Warnock, A.; Ivanov, V.Y.; Katopodes, N.D. Coupled modeling of hydrologic and hydrodynamic processes including overland and channel flow. *Adv. Water Resour.* **2012**, *37*, 104–126. [CrossRef]
18. Lerat, J.; Perrin, C.; Andréassian, V.; Loumagne, C.; Ribstein, P. Towards robust methods to couple lumped rainfall—Runoff models and hydraulic models: A sensitivity analysis on the Illinois River. *J. Hydrol.* **2012**, *418*, 123–135. [CrossRef]
19. Thompson, J.R. Simulation of Wetland Water-Level Manipulation Using Coupled Hydrological/Hydraulic Modeling. *Phys. Geogr.* **2004**, *25*, 39–67. [CrossRef]
20. Bellos, V.; Tsakiris, G. A hybrid method for flood simulation in small catchments combining hydrodynamic and hydrological techniques. *J. Hydrol.* **2016**, *540*, 331–339. [CrossRef]
21. Bout, B.; Jetten, V. The validity of flow approximations when simulating catchment-integrated flash floods. *J. Hydrol.* **2018**, *556*, 674–688. [CrossRef]
22. Costabile, P.; Costanzo, C.; De Bartolo, S.; Gangi, F.; Macchione, F.; Tomasicchio, G.R. Hydraulic Characterization of River Networks Based on Flow Patterns Simulated by 2-D Shallow Water Modeling: Scaling Properties, Multifractal Interpretation, and Perspectives for Channel Heads Detection. *Water Resour. Res.* **2019**, *55*, 7717–7752. [CrossRef]
23. Pato, J.F.; Caviedes-Voullième, D.; García-Navarro, P. Rainfall/runoff simulation with 2D full shallow water equations: Sensitivity analysis and calibration of infiltration parameters. *J. Hydrol.* **2016**, *536*, 496–513. [CrossRef]
24. Caviedes-Voullième, D.; García-Navarro, P.; Murillo, J. Influence of mesh structure on 2D full shallow water equations and SCS Curve Number simulation of rainfall/runoff events. *J. Hydrol.* **2012**, *448*, 39–59. [CrossRef]
25. Cea, L.; Bladé, E. A simple and efficient unstructured finite volume scheme for solving the shallow water equations in overland flow applications. *Water Resour. Res.* **2015**, *51*, 5464–5486. [CrossRef]
26. Arnold, J.G.; Srinivasan, R.; Muttiah, R.S.; Williams, J.R. Large area hydrologic modeling and assessment part I: Model development. *JAWRA J. Am. Water Resour. Assoc.* **1998**, *34*, 73–89. [CrossRef]
27. HEC. *HEC-RAS River Analysis System User's Manual Version 5.0*; Hydrologic Engineering Center: Davis, CA, USA, 2016.
28. Wilkison, D.H.; Armstrong, D.J.; Norman, R.D.; Poulton, B.C.; Furlong, E.T.; Zaugg, S.D. *Water Quality in the Blue River Basin, Kansas City Metropolitan Area, Missouri and Kansas, July 1998 to October 2004*; Scientific Investigations Report; US Geological Survey: Reston, VA, USA, 2006; Volume 5147, 170p.

29. Heimann, D.C.; Kelly, B.P.; Studley, S.E. *Flood-Inundation Maps and Wetland Restoration Suitability Index for the Blue River and Selected Tributaries, Kansas City, Missouri, and Vicinity, 2012*; Scientific Investigations Report; US Geological Survey: Reston, VA, USA, 2015; Volume 5180, 23p.

30. Missouri Department of Natural Resources. Water Protection Program: Total Maximum Daily Loads (TMDLs) for Blue River Jackson County, Missouri. Available online: https://dnr.mo.gov/env/wpp/tmdl/docs/0417-0418-0419-0421-blue-river-tmdl.pdf (accessed on 15 May 2020).

31. Amatya, D.M.; Jha, M.K.; Edwards, A.E.; Williams, T.M.; Hitchcock, D.R. SWAT-based phosphorus modeling of a karst watershed with an embayment, South Carolina. *J. Environ. Prot.* **2012**, *4*, 75–90. [CrossRef]

32. Schilling, K.E.; Gassman, P.W.; Kling, C.L.; Campbell, T.; Jha, M.K.; Wolter, C.F.; Arnold, J.G. The potential for agricultural land use change to reduce flood risk in a large watershed. *Hydrol. Process.* **2013**, *28*, 3314–3325. [CrossRef]

33. Burkart, C.S.; Jha, M.K. Site-specific simulation of nutrient control policies: Integrating economic and water quality effects. *J. Agric. Resour. Econ.* **2012**, *37*, 20–23.

34. Feng, H.; Jha, M.; Gassman, P. The Allocation of Nutrient Load Reduction across a Watershed: Assessing Delivery Coefficients as an Implementation Tool. *Rev. Agric. Econ.* **2009**, *31*, 183–204. [CrossRef]

35. Jha, M.K.; Gassman, P.W.; Arnold, J.G. Water Quality Modeling for the Raccoon River Watershed Using SWAT. *Trans. ASABE* **2007**, *50*, 479–493. [CrossRef]

36. Kling, C.L.; Panagopoulos, Y.; Rabotyagov, S.S.; Valcu, A.M.; Gassman, P.W.; Campbell, T.; White, M.J.; Arnold, J.; Srinivasan, R.; Jha, M.K.; et al. LUMINATE: Linking agricultural land use, local water quality and Gulf of Mexico hypoxia. *Eur. Rev. Agric. Econ.* **2014**, *41*, 431–459. [CrossRef]

37. Jha, M.K.; Pan, Z.; Takle, E.S.; Gu, R. Impact of climate change on stream flow in the upper Mississippi River Basin: A regional climate model perspective. *J. Geophys. Res.* **2004**, *109*, 1–12. [CrossRef]

38. Jha, M.; Arnold, J.G.; Gassman, P.W.; Giorgi, F.; Gu, R.R. Climate chhange sensitivity assessment on upper mississippi river basin streamflows using swat. *JAWRA J. Am. Water Resour. Assoc.* **2006**, *42*, 997–1015. [CrossRef]

39. Jha, M.K.; Gassman, P.W.; Panagopoulos, Y. Regional changes in nitrate loadings in the Upper Mississippi River Basin under predicted mid-century climate. *Reg. Environ. Chang.* **2013**, *15*, 449–460. [CrossRef]

40. Takle, E.S.; Jha, M.; Lu, E.; Arritt, R.W.; Gutowski, W.J. Streamflow in the upper Mississippi river basin as simulated by SWAT driven by 20th Century contemporary results of global climate models and NARCCAP regional climate models. *Meteorol. Z.* **2010**, *19*, 341–346. [CrossRef]

41. Lu, E.; Takle, E.S.; Manoj, J. The Relationships between Climatic and Hydrological Changes in the Upper Mississippi River Basin: A SWAT and Multi-GCM Study. *J. Hydrometeorol.* **2010**, *11*, 437–451. [CrossRef]

42. Jha, M.K.; Gassman, P.W. Changes in hydrology and streamflow as predicted by a modelling experiment forced with climate models. *Hydrol. Process.* **2014**, *28*, 2772–2781. [CrossRef]

43. Chattopadhyay, S.; Jha, M.K. Hydrologic response due to projected climatic variability in Haw River Watershed, North Carolina, USA. *Hydrol. Sci. J.* **2016**, *61*, 495–506. [CrossRef]

44. Abbaspour, K.C.; Yang, J.; Maximov, I.; Siber, R.; Bogner, K.; Mieleitner, J.; Zobrist, J.; Srinivasan, R. Modelling hydrology and water quality in the pre-alpine/alpine Thur watershed using SWAT. *J. Hydrol.* **2007**, *333*, 413–430. [CrossRef]

45. Moriasi, D.N.; Arnold, J.G.; Van Liew, M.W.; Bingner, R.L.; Harmel, R.D.; Veith, T.L. Model Evaluation Guidelines for Systematic Quantification of Accuracy in Watershed Simulations. *Trans. ASABE* **2007**, *50*, 885–900. [CrossRef]

46. Teng, J.; Jakeman, A.; Vaze, J.; Croke, B.F.; Dutta, D.; Kim, S. Flood inundation modelling: A review of methods, recent advances and uncertainty analysis. *Environ. Model. Softw.* **2017**, *90*, 201–216. [CrossRef]

47. Merwade, V.; Olivera, F.; Arabi, M.; Edleman, S. Uncertainty in Flood Inundation Mapping: Current Issues and Future Directions. *J. Hydrol. Eng.* **2008**, *13*, 608–620. [CrossRef]

48. De Moel, H.; Van Alphen, J.; Aerts, J.C.J.H. Flood maps in Europe—Methods, availability and use. *Nat. Hazards Earth Syst. Sci.* **2009**, *9*, 289–301. [CrossRef]

49. Bermúdez, M.; Zischg, A.P. Sensitivity of flood loss estimates to building representation and flow depth attribution methods in micro-scale flood modelling. *Nat. Hazards* **2018**, *92*, 1633–1648. [CrossRef]

50. Milanesi, L.; Pilotti, M.; Ranzi, R. A conceptual model of people's vulnerability to floods. *Water Resour. Res.* **2015**, *51*, 182–197. [CrossRef]

51. Arrighi, C.; Pregnolato, M.; Dawson, R.; Castelli, F. Preparedness against mobility disruption by floods. *Sci. Total. Environ.* **2018**, *654*, 1010–1022. [CrossRef]

52. Arrighi, C.; Oumeraci, H.; Castelli, F. Hydrodynamics of pedestrians' instability in floodwaters. *Hydrol. Earth Syst. Sci.* **2017**, *21*, 515–531. [CrossRef]
53. Costabile, P.; Costanzo, C.; De Lorenzo, G.; Macchione, F. Is local flood hazard assessment in urban areas significantly influenced by the physical complexity of the hydrodynamic inundation model? *J. Hydrol.* **2020**, *580*, 124231. [CrossRef]
54. Fu, J.-C.; Hsu, M.H.; Duann, Y. Development of roughness updating based on artificial neural network in a river hydraulic model for flash flood forecasting. *J. Earth Syst. Sci.* **2016**, *125*, 115–128. [CrossRef]
55. Pourali, S.H.; Arrowsmith, C.; Chrisman, N.; Matkan, A.A.; Mitchell, D. Topography Wetness Index Application in Flood-Risk-Based Land Use Planning. *Appl. Spat. Anal. Policy* **2014**, *9*, 39–54. [CrossRef]
56. Saksena, S.; Merwade, V. Incorporating the effect of DEM resolution and accuracy for improved flood inundation mapping. *J. Hydrol.* **2015**, *530*, 180–194. [CrossRef]
57. Liu, Z.; Merwade, V.; Jafarzadegan, K. Investigating the role of model structure and surface roughness in generating flood inundation extents using one- and two-dimensional hydraulic models. *J. Flood Risk Manag.* **2019**, *12*, e12347. [CrossRef]

MDPI
St. Alban-Anlage 66
4052 Basel
Switzerland
Tel. +41 61 683 77 34
Fax +41 61 302 89 18
www.mdpi.com

Water Editorial Office
E-mail: water@mdpi.com
www.mdpi.com/journal/water

www.ingramcontent.com/pod-product-compliance
Lightning Source LLC
LaVergne TN
LVHW070910170726
843515LV00005B/746